Islands & Rapids

A GEOLOGIC STORY OF HELLS CANYON

Islands & Rapids

A GEOLOGIC STORY OF HELLS CANYON

TRACY VALLIER

Confluence Press
Lewiston, Idaho

Copyright 1998 by Tracy L. Vallier

All rights reserved. Printed in the United States of America. No part of this book may be reproduced, stored in a retrieval system, or transmitted in any form or by any means—electronic, mechanical, photocopying, recording, or other—without written permission from the publisher. Inquiries should be addressed to Confluence Press, Lewis-Clark State College, 500 8th Avenue, Lewiston, Idaho 83501. Requests for reproduction or other uses of the figures and photographs should be addressed to the author at Confluence Press.

10 9 8 7 6 5 4 3 2

Book design by David C. Hoyt
Line drawings by Susan Vath and Gary Mann
Geologic map illustrations by Jeanette Smith
Back cover photograph by Mark E. LaMoreaux
Photographs by Tracy Vallier, except as noted
Glossary definitions, in part, from *Glossary of Geology,* (1997), American Geological Institute, Alexandria, VA

ISBN #1-881090-30-2

Library of Congress Catalog Number 98-070696

Distributed to the trade by
Midpoint Trade Books
1263 Southwest Boulevard
Kansas City, KS 66103

*For my mother, Pearl,
who hitched my wagon to a star*

Contents

Introduction to Islands and Rapids: A Geologic Story of Hells Canyon 1
    My Studies in Hells Canyon 2
    Wildlife, Spiders, and Insects 5
    Definition of Hells Canyon 6

Chapter 1. Geologic Time and Regional Geology 9
    Geologic Time 9
    Exotic Terrane 13
    Geologic History of the Blue Mountains Region 15

Chapter 2. Geology of Hells Canyon 23
    General Geology 23
    Snake River and the Cutting of Hells Canyon 24
    Pre-Cenozoic Rocks 27
        Stratified Rocks 28
        Intrusive Rocks 32
    Cenozoic Rocks 33
        Tertiary Gravels (pre-Columbia River Basalt Group) 33
        Columbia River Basalt Group 34
    Late Quaternary Deposits 35
        Landslide and Slump Deposits 36
        River Terraces 37
        Alluvial Fans 38
    Bonneville Flood 39
    Mazama Ash 41
    Rapids 41
    Shiny Rocks and Boulders 42

Chapter 3. Geologic Guide Between Hells Canyon Dam and the Mouth of the Grande Ronde River 45
    Introduction 45
    Segment 1   Hells Canyon Dam to Bernard Creek 46
    Segment 2   Bernard Creek to Sheep Creek 57
    Segment 3   Sheep Creek to Hominy Creek 63
    Segment 4   Hominy Creek to Upper Pittsburg Landing 67
    Segment 5   Upper Pittsburg Landing to West Creek 71
    Segment 6   West Creek to Getta Creek 78
    Segment 7   Getta Creek to Dug Bar 80
    Segment 8   Dug Bar to Cottonwood Creek 83
    Segment 9   Cottonwood Creek to China Garden Creek 97
    Segment 10 China Garden Creek to Grande Ronde River 101

Chapter 4. Geologic Guide Between Oxbow and Hells Canyon Dams 105
    Introduction 105
    Geology of the Oxbow Dam Area 105
        Oxbow Complex 111
    Guide to the Geology Along the Idaho Power Company Road 113
        0.0 to 21.6 Miles 113
        Hells Canyon Dam Area 126

Annotated Bibliography 129

Glossary 135

Care of the Canyon and Hiking Safety 145

Acknowledgments 149

Introduction to Islands and Rapids: A Geologic Story of Hells Canyon

The geology of Hells Canyon tells a story that includes both islands and rapids. To begin with, many of the rocks now exposed in Hells Canyon were once part of an island chain in the ancestral Pacific Ocean. This chain of islands probably existed far off the coast of the ancient North American continent. And rapids have played an important role in the creation of Hells Canyon. The pounding and swirling turbulence of the Snake River's effervescent rapids have continued to cut the canyon ever deeper even as relentless tectonic forces have lifted the surrounding landscape to ever greater heights.

Hells Canyon of the Snake River is dramatic and dynamic. Boldly forging a path through rocks that now form the border between Oregon and Idaho, the Snake River in Hells Canyon tenaciously incises its way into the continent. This deepening will end only when the river either ceases to flow through the canyon or when the tremendous tectonic forces no longer lift the surrounding landscape. The dams that impede the river's free flow in Hells Canyon are only slight perturbations in the history of the canyon. In a century or two the reservoirs behind those dams will be filled with millions of tons of sediment. Hopefully by that time other forms of energy will have replaced hydroelectric power. In either case, Brownlee, Oxbow, and Hells Canyon dams probably will be destroyed and the dormant river will awaken to restore a delicate balance between erosion and sedimentation. The river will wash away the sand and silt that will have filled those reservoirs, and golden beaches will once again line the water's edge (Figures 1-4).

Islands and Rapids tells a geologic story of Hells Canyon, and it includes a geologic guide that covers the canyon's entire length from the Oxbow Dam to the mouth of the Grande Ronde River: a distance of approximately 100 miles (Figure 5). The geology of Hells Canyon is complicated, yet by reading the introductory material in this book, some of the reference materials included in the annotated bibliography, and the glossary of geologic terms, even a beginning student of earth sciences can gain a significant understanding of the canyon's geologic evolution. Whether by foot or by boat, a trip through Hells Canyon can heighten our awareness of nature, increase our understanding of earth processes, influence our attitudes about this unique landscape, and demonstrate our need to preserve Hells Canyon for future generations.

The walls of Hells Canyon expose rocks that have a very exciting geo-

Figure 1. Anatahan Island, Commonwealth of The Northern Marianas Islands. Islands like Anatahan were common when the oldest rocks in Hells Canyon were young.
Nina Luttinger Photo

Figure 2. Wild Sheep Rapids in Hells Canyon. Turbulent waters of the Snake River will continue to deepen Hells Canyon as tectonic forces lift the surrounding landscape.

April 1963. Based on a suggestion by Professor William Taubeneck, Bruce Nolf and I journeyed across Oregon to examine rocks in Hells Canyon for a possible dissertation topic. The Idaho Power Company road to the Hells Canyon dam site was being constructed at the time and Bruce and I drove out as far as we could to look at the rocks. At every stop I furrowed my brow and scratched my head. I had no idea about the origin of these rocks. Between 1963 and 1966, eleven months of field work and the investigation of hundreds of rock thin sections were required to begin understanding the geologic mysteries of Hells Canyon. Little did I know that these rocks would evoke a scientific passion in me that would last for more than thirty-five years.

Figure 3. Hells Canyon looking north toward Barton Heights and Bull Creek. This is one of the most rugged parts of Hells Canyon (*Book frontispiece*).

Figure 4. Hells Canyon as seen from the north side of Dry Diggins Ridge in Idaho. The Snake River flows more than a mile in elevation below Dry Diggins Ridge (*Color section, page 89*).

logic history. The canyon has been cut through an exotic terrane that formed in the ancestral Pacific Ocean, moved across part of that ocean under the relentless force of plate tectonics, and subsequently was welded to the ancient North American continent.

Interpretations of the canyon's exciting geologic history are based on the work of many scientists, students, and others who have chosen to question the origin and evolution of the rocks. In time, many of these interpretations will change because geology is a vigorous science that has sustained several new and important theories and models in recent decades. In fact, some of the interpretations that I present in this book will no doubt undergo revision as scientists collect more data and modify the current geologic models.

MY STUDIES IN HELLS CANYON

I deeply respect and care about Hells Canyon. It has been my mentor and comrade for more than thirty-five years. The river and rugged canyon walls influence and challenge my thoughts much like a demanding professor challenges the wit and wisdom of his or her students. Within the dynamic setting of the Hells Canyon region I test both my physical and mental strengths. I revel at the moods of the Snake River, witnessing its angry churning during floods and feeling its soothing caress when I float the peaceful waters of backeddies. I have spent many hypnotic hours watching the wild water as it explodes through raging rapids. And seemingly endless hours have been spent with a four-pound hammer in my grasp, while I fervently questioned the origin and history of the rocks.

I began my studies of the Hells Canyon region in 1963. My work has involved quests—geologic and personal journeys that I share throughout the text as well as in the sidebars and in several of my favorite photographs. Although I spent most of my first field season in the southeastern Wallowa Mountains near Fish Lake, even then I was targeting my mapping efforts toward Hells Canyon. I have devoted a large part of my life to conducting field work in the region, and then examining data in numerous laboratories, drafting figures, compiling maps, and writing papers and reports in stale offices.

While working on my dissertation in the early 1960s, the theories and tectonic models that I had learned in college courses and from reading geologic literature influenced my mapping and interpretations. Soon, however, a new hypothesis, mostly untested at that time, began reaching the pages of scientific journals. After conclusive testing this hypothesis (called sea floor spreading) led to the *theory of plate tectonics*. Nearly all earth scientists now accept the theory of plate tectonics, which serves as a model for interpretations in many facets of earth history. In fact, the theory is so widely accepted that it has become the *plate tectonic paradigm*: an all-inclusive model used to explain the dynamics of the earth's crust and upper mantle. My interpretations of the geology of the Hells Canyon region are strongly influenced by the plate tectonic paradigm. A new paradigm may yet influence the interpretations of my successors.

An important axiom in geology is *uniformitarianism*, which holds that "the present is the key to the past." This axiom has been very significant for understanding Hells Canyon geology. For example, between 1976 and 1988 I was involved in offshore and onshore studies of the Aleutian, Tonga, and Marianas **island arcs** (bold-face words are defined in the glossary) while I

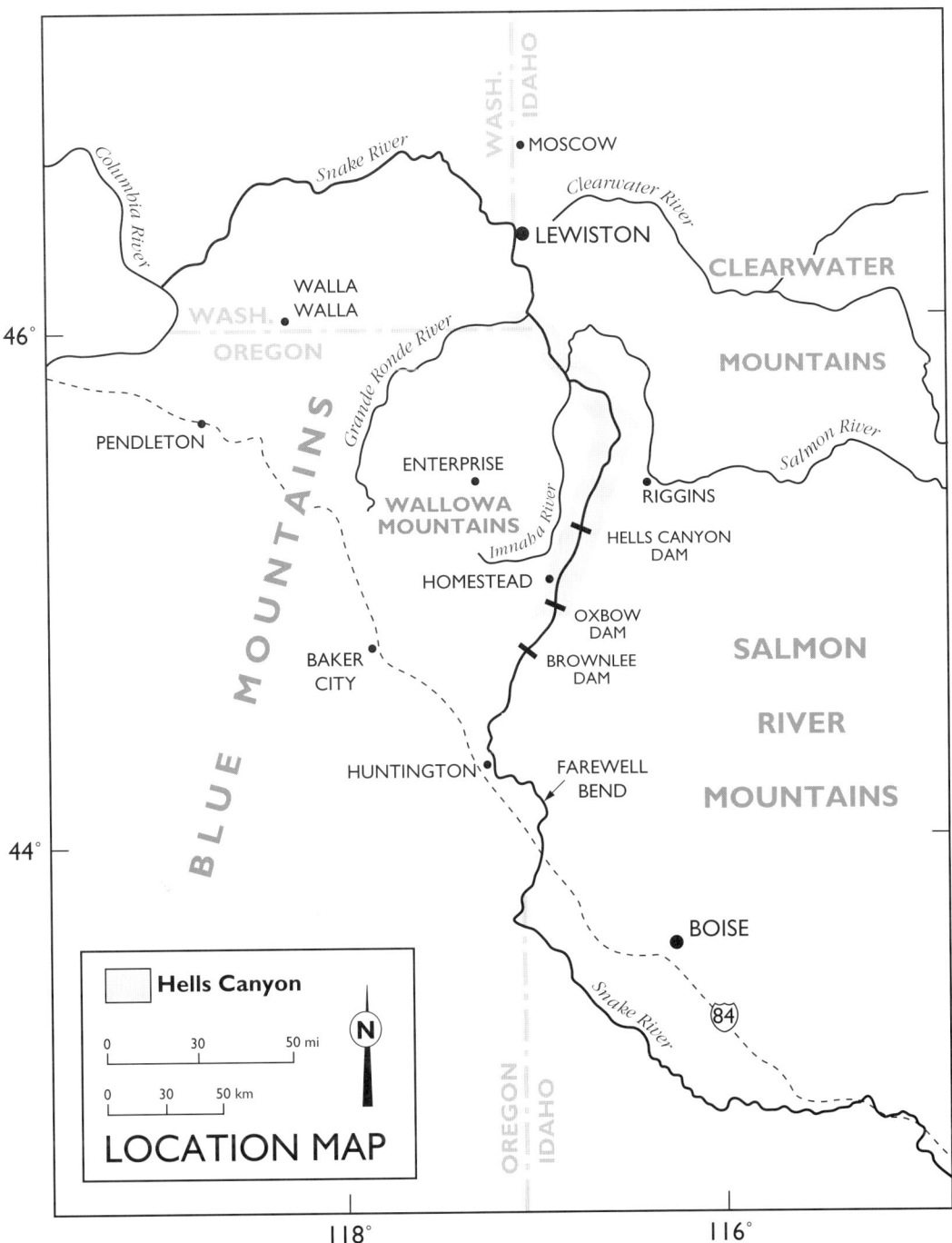

Figure 5. Location map of Oregon, Idaho, and Washington showing the area (shaded) discussed in this book.

worked as a marine geologist for the U. S. Geological Survey. Early in those studies it became apparent to me, and to others, that the rocks in Hells Canyon are remarkably similar to those rocks that I dredged from the insular slopes of the island arcs and collected from outcrops on the Pacific islands. Laboratory data confirmed the field work. Consequently, an understanding of the rocks and active geologic processes in modern island arcs greatly influenced my interpretations of Hells Canyon geology.

Figure 6. Rattlesnakes are common in Hells Canyon. Most are shy and very docile, but take no chances with them. Snakes are important in Hells Canyon ecosystems and should not be destroyed.

Figure 7. Mountain goats may greet the tired traveler high above the Snake River near the abandoned Dry Diggins fire lookout.

Figure 8. Spiders are prolific along trails in the late summer. Most are harmless. Many are brilliantly colored.

WILDLIFE, SPIDERS, AND INSECTS

Although this book primarily addresses rocks and geologic processes, no naturalist can overlook the thrill of unpredictable encounters with animals, the intrigue of observing spiders and insects, and the sensation of exploring a vast array of plant life (Figures 6-8). The histories of the Native Americans and early explorers, of ranchers, gold miners, and dam builders, have also fueled my imagination; so have the discoveries of house pits, arrowheads, grave markers, rusted shovels, mine tunnels, and dilapidated cabins (Figures 9-13). The personal experiences have made my field work in Hells Canyon more than an odyssey into rock history; they have widened a road to include a journey into a broader realm of the natural sciences and thereby contributed to my better understanding of (and respect for) all canyon inhabitants—past and present.

The presence of wildlife makes working in the canyon particularly rewarding: I've experienced the intensity of coming face to face with bears, of being stalked by mountain lions, of swimming with river otters, and of encountering dozens of rattlesnakes. How can I forget that massive bull elk, who stood boldly in my path while his cows and calves sauntered over the ridge to presumed safety? Or the sunrise terror of finding a rattlesnake coiled near my sleeping bag? I still sense the rush of air through angry wings and hear the piercing screams of two bald eagles who flared their outstretched talons when I approached the Ponderosa pine tree where their eaglets nested. I remember a rocky ledge above Eagle Bar when, for several minutes, a brilliant hummingbird tried to pluck nectar from my new red vest.

At times I also recall the way my four-pound single-jack hammer arced through the morning sky for a rendezvous with a rattlesnake's head twenty yards away; I still feel pangs of remorse because I inadvertently had killed a creature of no immediate threat to me. At still other times I remember the taste of boiled grouse in a remote camp high above the river—a remarkable dinner that was provided with the help of a well-aimed rock.

One morning during a snack break, as my colleagues and I explored steep parts of Three Creeks Canyon, a cinnamon-colored bear grew curious about the peanuts and raisins we were eating. I chased it away by swinging

Figure 9. Horse Heaven cabin, maintained by the U.S. Forest Service, sits next to the ridge trail that leads south to Stevens Saddle. High peaks of the Seven Devils Mountains decorate the north skyline.

Figure 10. (opposing page) Oxbow Dam. Hells canyon was changed by the construction of the Brownlee, Oxbow, and Hells Canyon dams during the late 1950s and 1960s. The dams are temporary features, however, when compared to the age of the rocks that form their solid buttresses.

Figure 11. Dave Fredley on the remains of an automobile, Sheep Creek alluvial fan, 1971. These relics have been removed by over-zealous collectors *(Color section, page 89).*

Figure 12. Abandoned cabin along Sluice Creek *(Color section, page 90).*

Figure 13. High above Three Creeks, hikers gaze into rugged Hells Canyon *(Color section, page 90).*

Figure 14. Far below the trail leading to Hat Point, the Snake River winds toward Sheep Creek *(Color section, page 91).*

my hammer threateningly and shouting expletives that I can't repeat here. I still chuckle whenever I remember a saucy mother mink and her babies who ran through the rocks and squealed while trying to steal a trout as it dangled from the end of my son's fishing line. Near Dry Diggins Fire Lookout a wild goat and her kid once amused me by eating trail mix from my outstretched hand.

The depths of the canyon are too hot and dry for mosquitoes, but grasshoppers and spiders flourish. They reflect a delicate balance within some of the canyon's ecosystems. During years of prolific overpopulation, grasshoppers eat most of the tender vegetation. Furthermore, they create hazards for hikers who wander the steep canyon slopes. It is a harrowing experience to walk uphill into a shower of grasshoppers that are merely following the pull of gravity toward lower parts of the canyon; they can inflict painful welts on unprotected skin. Indeed, the energies of jumping and flying grasshoppers, both kinetic and potential, are so great that they can dislodge loose sand grains and pebbles where they land. For example, I recall measuring the twenty-foot distance that a small pebble, about a centimeter in diameter, flew after struck by a grasshopper of average size. No doubt, grasshoppers are important erosive agents along the steep slopes of Hells Canyon.

I often wonder how much of Hells Canyon was eroded by energetic grasshoppers.

Although I enjoy watching the many varieties of spiders and studying their intricate webs, I caution all canyon visitors to be wary of the brown recluse and black widow spiders. In 1991 a brown recluse bit my brother, Kent, several times while he dozed in a sleeping bag; the bites required a doctor's attention. Most spiders, however, pose no danger to the hiker. Multicolored spiders build webs across river trails to catch unwary grasshoppers and other insects. The webs can be a nuisance to the hiker. In 1979 I noticed that the Idaho side of the Snake River south of Pittsburg Landing had only two major varieties of spiders, whereas the Oregon side had six. Is the river a major barrier?

As for scorpions, I have seen only one. After spending the night camped on a sandbar near the mouth of the Salmon River, I shook a scorpion out of my boot the following morning.

DEFINITION OF HELLS CANYON

Both the name and exact geographic location of Hells Canyon are controversial. The Snake River Canyon between Oregon and Idaho has been called the Grand Canyon of the Snake, Box Canyon, and Hells Canyon. Hells Canyon was named for Hells Canyon Creek, which enters the river at the visitor's center just below Hells Canyon Dam. In her book, *My Home Below Hells Canyon*, Grace Jordan implies that Hells Canyon is a short segment of the canyon above her former home at Kirkwood Creek. That definition is too restrictive. Likewise, many river guides would like visitors to believe that they can see Hells Canyon only by either floating through it below Hells Canyon Dam or by jetting through the rapids into the deepest part of the canyon. Some of the more rugged and spectacular parts of the canyon occur along the reservoir south of the Hells Canyon Dam and these can be seen by automobile.

I refuse to be caught in this controversy, so I'll define Hells Canyon solely for purposes of this book. I consider Hells Canyon to be that stretch of the

Snake River Canyon between the Oxbow and the mouth of the Grande Ronde River. This is a liberal definition, but it is logical because of the canyon's geologic history. If I were to base my definition only on physiography, I would limit Hells Canyon to that segment of canyon between Kinney and Sheep creeks (Figure 14), the area that more or less parallels the Seven Devils Mountains. Hells Canyon Dam is an artificial and ephemeral boundary and should not be used in the definition.

Is Hells Canyon the deepest river canyon in the world? Of course not. Is it the deepest in North America? It is, if the depth is measured only from the eastern (Idaho) side where one segment of the canyon is more than 8,000 feet deep. If the depth of the river canyon is measured from the Oregon side near Hat Point, then the canyon is approximately a hundred feet deeper than the Grand Canyon of the Colorado River. Where the height of canyon walls are jagged, uneven, and asymmetric, like those in Hells Canyon, then perhaps the depth of a canyon should be measured from its lower side along any stretch of a river. If so, then Hells Canyon is deeper near Hat Point than the Grand Canyon, but not as deep as other river canyons in North America. For example, a fork of the Salmon River in Idaho has cut a canyon that may be deeper, and parts of the Kings River Canyon in California in places probably are deeper.

CHAPTER 1

Geologic Time and Regional Geology

GEOLOGIC TIME

Time has a dramatic impact on our lives. The savage ring of an alarm clock awakens us and deadlines plague our very existence. And time can run out on us at the whim of an unforeseen event. As humans, we generally measure time in centuries, decades, years, months, weeks, hours, minutes, and seconds. Based on our limited experience, it is difficult to think in terms of thousands, millions, and billions of years. Like so many others, I believed as a youth that a Superior Being created the earth and that it was only about 6,000 years old. Using a model of time that is laden with religious contexts and the experience of a short human life span, I had to exert considerable effort to comprehend time spans of a million years, 100 million years, and a billion years.

Nevertheless, scientists using radiometric dating techniques have provided strong evidence that the earth has persisted for approximately 4.5 billion (4,500 million) years. Simple life forms began inhabiting the earth about 2.5 billion years ago. Complex forms have been living for about 600 million years. The oldest rocks in Hells Canyon are at least 300 million years old and, as a deep Snake River gorge, Hells Canyon has existed for at least 2 million years. Small, ape-like creatures apparently roamed the earth 6 million years ago. Hominids with stone tool technology existed more than 2 million years ago. Humans similar to us have been tramping across the continents for perhaps 200,000 years. People have populated North America for at least 12,000 years. Few individual people, however, have had life spans greater than 100 years.

What is geologic time?

No one really understands, nor can anyone adequately explain, geologic time. Like the smallness of an atom or the immense distance of a light year, the concept of geologic time is nearly incomprehensible. How can one comprehend the length of one million years? What does it mean to say, "These rocks were flowing rivers of lava 235 million years ago?" Well, I guess one could say that 235 million years is a lot more than 10 million years and a lot less than 1,000 million years—or the age of the Earth at 4,500 million years.

Rocks—those apparently sedentary objects—can speak to us if we listen. They can tell us about geologic time. For example, how long does it take for a rock to break down to sand-sized grains and travel to the ocean floor by way of streams, ice, wind, and ocean currents? How long does it take a boulder of **basalt** to weather, and for all its component parts to be transported to the ocean, if we place the boulder on a ridge between He Devil and She Devil mountains? In Wild Sheep Rapids? In a meadow near Pittsburg Landing? Would it take a thousand years? One hundred thousand years? A million years?

For that matter, how long does it take to turn dead animals and dead plants into fossils (Figures 15 and 16)? How long does it take to turn mud into rock?

Perhaps I can put geologic time into perspective. Let us accept the premise that the earth is 4,500 million years old, that simple life began at least 2,500 million years ago (Ma), and that complex organisms were in the oceans by 600 Ma. Sharks were abundant around 400 Ma; dinosaurs roamed the earth from about 230 to 65 Ma; small horses were galloping across the Great Plains of North America at 50 Ma; human-like apes appeared as recently as 6 Ma; and humans—about as we know them—have existed for the last 0.2 million (200,000) years. From this perspective, then, a human's average life span of 70 to 80 years is minuscule; it is almost insignificant.

I like to compare geologic time to a twenty-four hour clock, with the origin of the earth (about 4,500 Ma) starting at midnight (00:00 hours). When the earth is 2,500 million years old and when primitive life first appears, thirteen and one-third hours have already passed and it is 1:20 p.m. The minutes and seconds tick on for another 1,400 million years as the earth and its organisms evolve. When complex animals become abundant about 600 Ma, nearly 21 hours have passed and afternoon has turned into evening; in fact, it is 8:40 p.m., a little more than 3 hours before midnight (present). Sharks are swimming in the oceans by 10:00 p.m. and small dinosaurs are walking the earth by 10:44 p.m.; dinosaurs become extinct at 11:39, just 21 minutes before midnight. Furry, manlike apes probably walked the earth by 11:58:40, slightly more than a minute before midnight. Approximately 0.00025 seconds ago, a 70 year-old person was born.

According to this twenty-four hour clock, the oldest rocks in Hells Canyon formed on brand new oceanic crust about 300 Ma (at approximately 10:24 p.m.). The thick sequence of lava flows that forms the Columbia River

Figure 15. Fossil productids (*Megousia*) in the Permian Hunsaker Creek Formation. This outcrop is near Homestead Creek in southern Hells canyon.

Figure 16. Fossil fern branch in the Jurassic Coon Hollow Formation at Pittsburg Landing.

Subdivisions of Geologic Time and Symbols

ERA	PERIOD AND SUBPERIOD		EPOCH	AGE (Ma)
CENOZOIC	QUATERNARY (Q)		Holocene	0.01
			Pleistocene	1.8
	TERTIARY (T)	NEOGENE SUBPERIOD	Pliocene	5.3
			Miocene	23.7
		PALEOGENE SUBPERIOD	Oligocene	36.6
			Eocene	57.8
			Paleocene	66.4
MESOZOIC	CRETACEOUS (K)		Late	
			Early	144
	JURASSIC (J)		Late	
			Middle	
			Early	208
	TRIASSIC (TR)		Late	
			Middle	
			Early	245
PALEOZOIC	PERMIAN (P)		Late	
			Early	286
	PENNSYLVANIAN (P)		Late	
			Middle	
			Early	320
	MISSISSIPPIAN (M)		Late	
			Early	360
	DEVONIAN (D)		Late	
			Middle	
			Early	408
	SILURIAN (S)		Late	
			Middle	
			Early	438
	ORDOVICIAN (O)		Late	
			Middle	
			Early	505
	CAMBRIAN (Є)		Late	
			Middle	
			Early	570
PROTEROZOIC	NONE DEFINED			2500
ARCHEAN	NONE DEFINED			3800

Figure 17. Geologic time scale (modified from the time scale used by the U.S. Geological Survey and the time scale that was published in 1983 for the *Decade of North American Geology*). The age column is millions of years ago (Ma). Rocks older than 570 Ma are also referred to as Precambrian. Pre-Archean (not shown) is an informal time term without a specific rank that has an age range between 4,500 and 3,800 Ma.

Basalt Group flowed out of fissures beginning 17 Ma or at about 11:55 p.m. Hells Canyon of the Snake River, mostly cut within the past 2 million years, records less than a minute on the twenty-four hour clock—a very short segment in the earth's evolution.

How do scientists measure geologic time? Geologic time is measured by both *relative* and *absolute* methods. *Relative* time relies on fossils and basic geologic principles; *absolute* time relies on the radioactive decay of elements—the ticking of natural atomic clocks. From studies using both relative and absolute time, geologists have constructed a geologic time scale (Figure 17). We need to be aware that the ages assigned to this time scale will change some as new data become available and also that the uncertainty of the assigned ages increases with the length of geologic time. In other words, the greater the length of geologic time, the greater the uncertainty of the age that is given in Figure 17.

Several axioms guide geologists as they place rocks within *relative* time sequences. For example, in a sequence of sedimentary rocks the youngest rocks are at the top—unless the sequence is overturned. Younger rocks (such as a granite **dike**) cut across older rocks. Furthermore, the same assemblage of fossils from different sedimentary rock units means that the rocks are the same approximate age. These simple axioms are very important in determining relative ages. In Hells Canyon, the flat-lying Columbia River Basalt flows are younger than the rocks they cover. A dike of **quartz diorite** that cuts **gabbro** near Dug Bar is younger than the gabbro. All of the rocks that contain the fossil clam, *Daonella beedi*, are of Middle Triassic age.

During the last 2 centuries, geologists have been placing sedimentary rocks in some semblance of order by using fossils to find strata of the same age. This technique is called *correlation of strata*. The discovery of radioactivity and the techniques used to measure it, however, have permitted the assignment of *radiometric* ages to the rocks. When a fossil assemblage is found (or sometimes just one important and diagnostic fossil), geologists are now able to assign a more precise age to the rock than was possible in the past. For example, we can say that all rocks containing the fossil clam *Daonella beedi* are about 235 to 230 million years old.

Several methods are used in *geochronometry*, the measurement of mineral and rock ages using radiometric clocks. In most instances, these mineral and rock ages are determined by the breakdown of radioactive isotopes. Isotopes of an element have the same number of protons in the nucleus of the atom but a different number of neutrons. For example, uranium has three naturally occurring isotopes: ^{238}U, ^{235}U, and ^{234}U. All three of these uranium isotopes are radioactive. The radioactive decay of ^{238}U (*parent*) gives rise to a *uranium series* that includes several other isotopes as intermediate *daughters* and ends in a stable lead isotope (^{206}Pb). By measuring the relative amounts of the radioactive *parent* atom (^{238}U) and the radiogenic *daughter* (^{206}Pb) through the use of a mass spectrometer, and by knowing decay constants and half-life (the time it takes for half the parent to decay to other isotopes), a *radiometric age* is determined. Other elements with parent isotopes that decay naturally (and thereby can be used for geochronometry) include thorium, rubidium, carbon, strontium, and potassium.

Many of the igneous rocks in Hells Canyon have been dated using radioactive isotopes, particularly those of uranium and potassium.

Some European epoch names are used in this book because they are so common in geologic literature. In the Triassic, these are Ladinian (Middle

Triassic), Karnian (early Late Triassic), and Norian (late Late Triassic). In the Jurassic, the epoch names used are Bajocian (early Middle Jurassic), Callovian (late Middle Jurassic) and Oxfordian (Late Jurassic).

EXOTIC TERRANE

Most of Hells Canyon is eroded through an *exotic terrane* that began its history as an island arc (figure 18). The older rocks that crop out beneath layered lava flows of the Columbia River Basalt Group originated elsewhere as islands and parts of adjacent ocean floors, perhaps far out in the ancestral Pacific Ocean. They were subsequently added (accreted) to the North American continent. Rock characteristics, rock ages, and paleomagnetic data all indicate an exotic origin for these older rocks. The exotic terrane story is complex, and it challenges our imagination, but I believe it also gives us a greater appreciation for the geologic story of Hells Canyon.

Exotic terrane is an abbreviated name for a **tectonostratigraphic terrane** that did not form where it is presently found. In fact, it may have traveled a long distance across the surface of the earth by plate tectonic processes to reach its present location. Tectonostratigraphic terrane is defined as a **fault**-bounded geologic entity of regional extent characterized by a geologic history distinct from that of neighboring terranes. The tectonostratigraphic terrane concept has been used by earth scientists since the late 1970s and is an important working hypothesis for interpreting the geologic history of much of western North America and of other circum-Pacific countries. Where several terranes have amalgamated (joined together) prior to accretion, the entire mass is referred to as a composite terrane. If the adjective "exotic" is used before the noun "terrane," earth scientists presume that the terrane has undergone extensive lateral movement across the surface of the earth. For a better understanding of Hells Canyon in the context of an exotic terrane origin, I recommend that you watch a 28-minute video titled "Exotic Terrane." (See the annotated bibliography near the end of this book.)

Figure 18. Sketch of an island arc showing the **subduction zone**, fore-arc and back-arc regions, and the volcanic (magmatic) axis. There is a great deal of vertical exaggeration. Most volcanic and **plutonic** rocks in Hells Canyon, particularly those older than the Martin Bridge Limestone, formed near the volcanic axis of an island arc. The subduction zone and outer parts of the fore-arc region represent the Baker and Grindstone terranes, whereas the arc massif is characteristic of the Wallowa and Olds Ferry terranes (Figure 19).

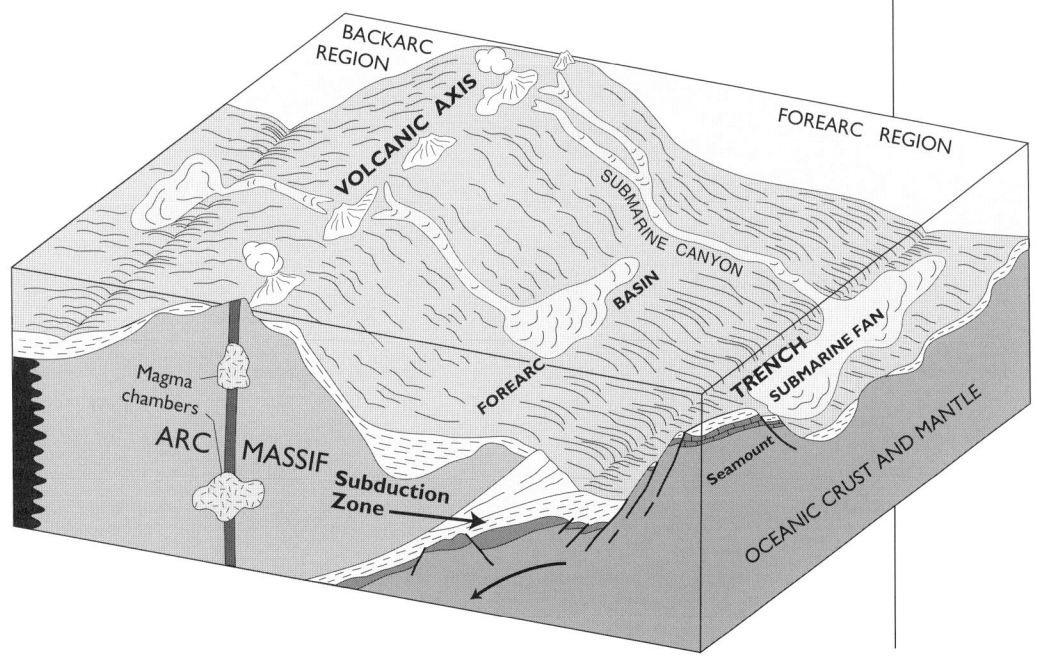

14 ISLANDS AND RAPIDS

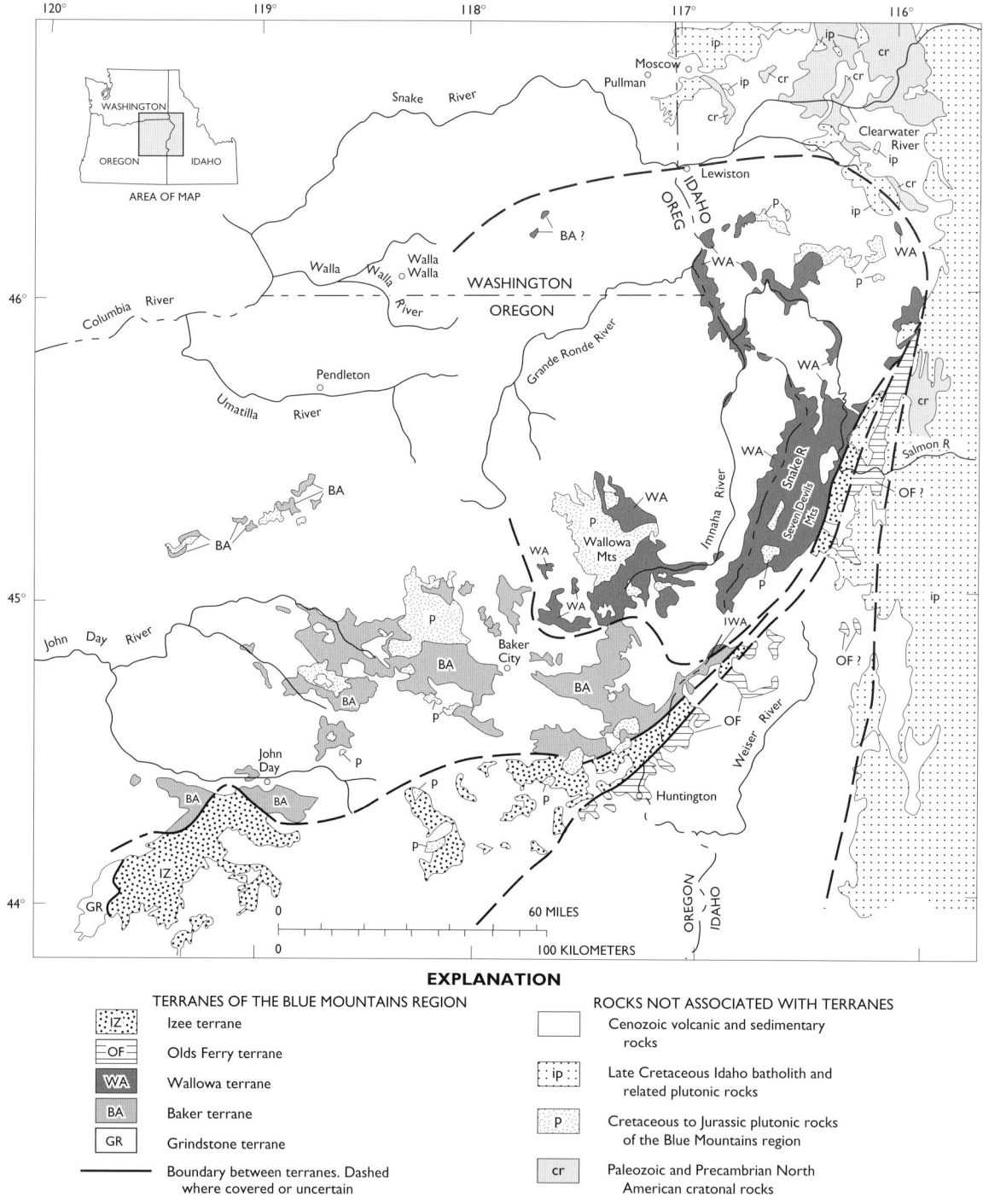

Figure 19. Terranes of the Blue Mountains Island Arc (from Vallier, 1995). Paleomagnetic studies indicate that the terranes were rotated about 60 degrees in a clockwise sense during the Late Cretaceous. By rotating the terranes back 60 degrees, it is apparent that the axes of at least the Olds Ferry and Izee terranes initially paralleled the border of the ancient North American continent.

The video is available for purchase through Confluence Press and at visitors' centers in the Wallowa-Whitman National Forest.

The older (pre-Cenozoic or pre-Tertiary) rocks of the Blue Mountains province in eastern Oregon, western Idaho, and southeastern Washington comprise a composite exotic terrane that formed in the ancestral Pacific Ocean. This composite terrane—or superterrane—is referred to as the Blue Mountains Island Arc (Figure 19). The Blue Mountains Island Arc formed near and along ancient subduction zones in a setting similar to the present-day "ring-of-fire" island arcs that border the western and northern margins of the Pacific Ocean. The Aleutian, Tonga, Kurile, and Mariana island arcs are modern examples of the Blue Mountains Island Arc.

Rocks that formed on islands and seamounts in the Blue Mountains Island Arc traveled hundreds of miles on the back of one or more tectonic plates in the ancient Pacific Ocean. The rocks subsequently were wrapped in the embrace of a more ancient continent, and are now stroked by rapids of the Snake River.

The Blue Mountains Island Arc consists of five separate terranes that formed within different parts of the arc: Baker, Grindstone, Izee, Olds Ferry, and Wallowa terranes (Figure 19). The Wallowa and Olds Ferry terranes formed along volcanic axes of the arc. The axis of volcanism shifted from the Wallowa Terrane to the Olds Ferry Terrane about 225 to 220 Ma when a probable change occurred in the relative convergence directions (and possibly velocities) of the oceanic plates. The Baker and Grindstone terranes, geologically more complex than the Wallowa and Olds Ferry terranes, formed between the volcanic axis and trench in an area broadly referred to as a fore-arc region—that includes the subduction zone. The Izee Terrane is composed mostly of sedimentary rocks (sandstone and siltstone). Sediments in the Izee Terrane were eroded from adjacent older terranes and deposited in a deep basin that developed within the fore-arc region of the Olds Ferry volcanic islands.

The pre-Cenozoic rocks that crop out beneath the younger lavas in Hells Canyon, from the Oxbow to the mouth of the Grande Ronde River, are part of the Wallowa Terrane. Rocks in the Wallowa Terrane formed on and near islands and seamounts along, and close to, the volcanic axis of the Blue Mountains Island Arc.

GEOLOGIC HISTORY OF THE BLUE MOUNTAINS REGION

My current interpretations of the pre-Cenozoic geologic history of the Blue Mountains, greatly simplified, are shown in Figure 20 (A-G). The sketches show the development of the Blue Mountains Island Arc in seven stages that span the Late Devonian through most of the Early Cretaceous, a time interval of about 250 million years. The interpretations for the first 100 million years, from the Late Devonian to Early Permian, are very speculative. I feel much more confident about interpretations for the time interval between about 270 and 120 Ma. The history culminates in the accretion of the island arc to ancient North America. The present boundary between the Blue Mountains Island Arc and the ancient North American continent is only a few miles wide and can be observed in many outcrops along a narrow, north-south strip between the towns of McCall and Orofino, Idaho. It is well exposed near Riggins, Idaho.

16 ISLANDS AND RAPIDS

A   Late Devonian(?) to Early Permian (370?-270 Ma)

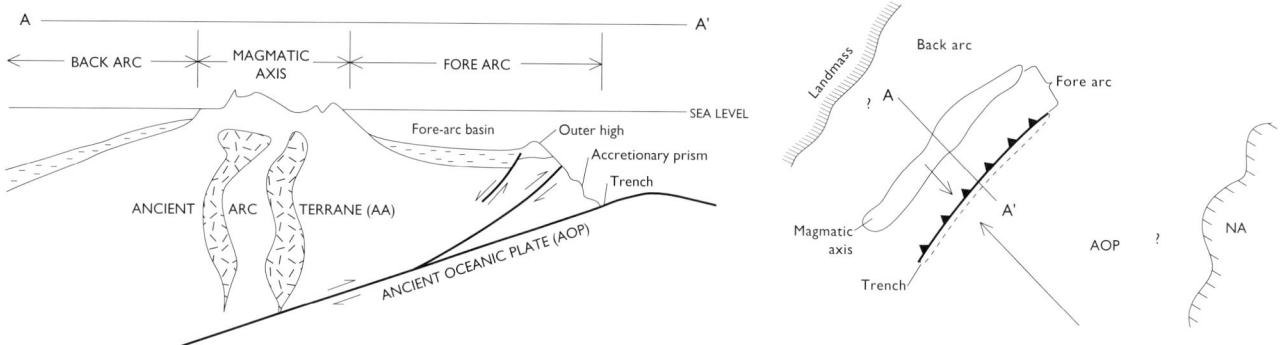

Figure 20. Seven geologic sketches (A-G) showing the evolution of the Blue Mountains Island Arc in the Late Devonian through Early Cretaceous interval. These diagrams are from Vallier (1995). Abbreviations are the following: AA, ancient arc terrane; AOP, ancient oceanic plate; BA, Baker Terrane; IZ, Izee Terrane; NA, North American continent before accretion of the Blue Mountains Island Arc; OF, Olds Ferry Terrane; OT, other terranes; PP, Pacific (and Farallon) Plate; WA, Wallowa Terrane.

20A. Late Devonian to Early Permian (370?-270 Ma): Data are scant for this time interval and my interpretations are very speculative. The explanation for this sketch, however, should be used for the next 6 sketches. The cross section (from A to A' on the small map) is somewhat self-explanatory. It is a vertical drawing of the island arc along the line shown on the small map to the right of the cross section. The small map shows, from lower right to upper left, a small part of the North American continent (NA), an ancient oceanic plate (AOP), an arrow showing the direction of sea-floor movement, a trench, the subduction zone with an **accretionary prism** shown by triangular teeth (teeth on subduction zones and **thrust faults** are placed on the over-riding plate and hanging wall, respectively) on a line that parallels the magmatic axis (also referred to as the volcanic axis; the site where volcanoes grew and beneath which **plutons** crystallized), a fore-arc region that probably contained a basin, an arrow showing that the magmatic axis was moving oceanward with relation to the oceanic plate (referred to as convergence of plates), and the volcanic or magmatic axis. I show a landmass to the northwest (upper left) because it is possible that part of what is now northern Asia, or perhaps just another exotic terrane, existed to the northwest of the Blue Mountains Island Arc during the Devonian to Early Permian interval.

20B. Early Permian (270-260 Ma): This was the first of two very active times of volcanism and tectonism in the evolution of the Blue Mountains Island Arc, and particularly of the Hells Canyon region (Wallowa Terrane). The subduction zone was piling up debris, including fragments of seamounts, from the deep ocean floor along the inner trench wall in the fore-arc region (Baker Terrane) and silica-rich volcanoes were throwing rhyolitic **pyroclastic** materials into the atmosphere and down the slopes of the volcanoes (Wallowa Terrane). We have no idea what the ocean floor was like between the island arc and the North American continent at this and other times during the arc's evolution. However, geologists believe that many of what are now individual continents were grouped together in a supercontinent called Pangea during the Permian. Paleomagnetic data from Permian rocks in Hells Canyon strongly suggest that the Wallowa Terrane was located at about 35° to 25° North Latitude, a location that was north of its position later on in the Middle and Late Triassic. Furthermore, marine fossils from the Permian rocks suggest that the animals lived in colder waters than did the Triassic animals. We don't know what constituted the pedestal or basement upon which the

B Early Permian (270-260 Ma)

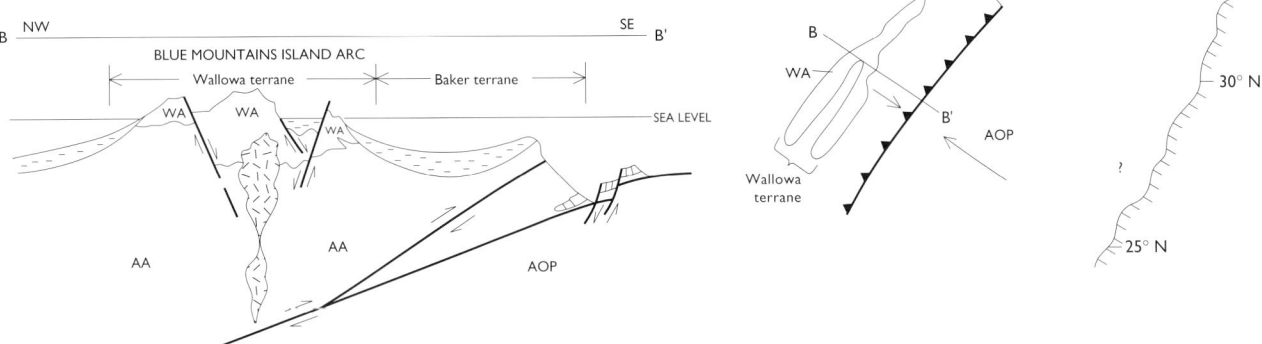

C Late Permian and Early Triassic (260-235 Ma)

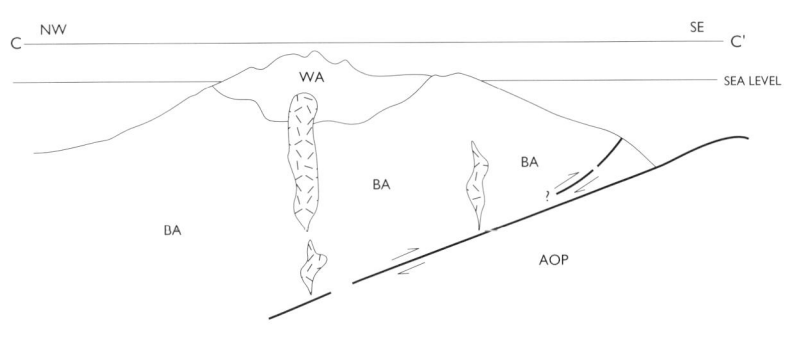
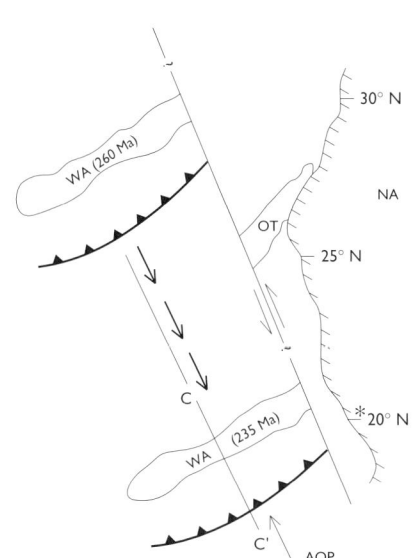

Permian volcanoes were built. A probable Late Pennsylvanian (about 309 Ma) radiometric age of a rock from the Cougar Creek Complex suggests that the Permian rocks of the Wallowa Terrane formed on an older arc (AA in the figure) that may be as old as Late Devonian in age. This presumption is speculative and more work is needed to find the oldest rocks in the Wallowa Terrane.

20C. Late Permian and Early Triassic (260-235 Ma): The Wallowa Terrane was mostly quiet volcanically during this time interval. Some plutons crystallized, but there is no evidence for prolific volcanic activity. The important process was translation of the island arc southward relative to ancient North America, probably along **strike-slip** or **transform faults** in the oceanic plates. The small map shows this translation of the island arc from northwest to southeast. I speculate that another terrane (OT for other terrane), already attached to North America, may have been left behind during this translation. Perhaps it resides in British Columbia. From the Early Triassic until final accretion during the Early Cretaceous, the island arc and North American continent moved northward together. But how far was the island arc from the west coast of the North American continent? An asterisk (*) shows the location where the island arc will ultimately collide with North America.

D Middle and early Late Triassic (235-225 Ma)

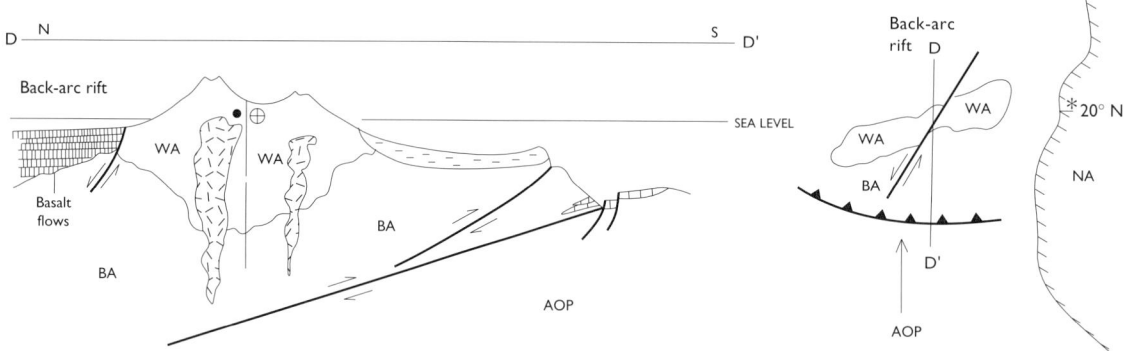

E Late Triassic and Early Jurassic (225-190 Ma)

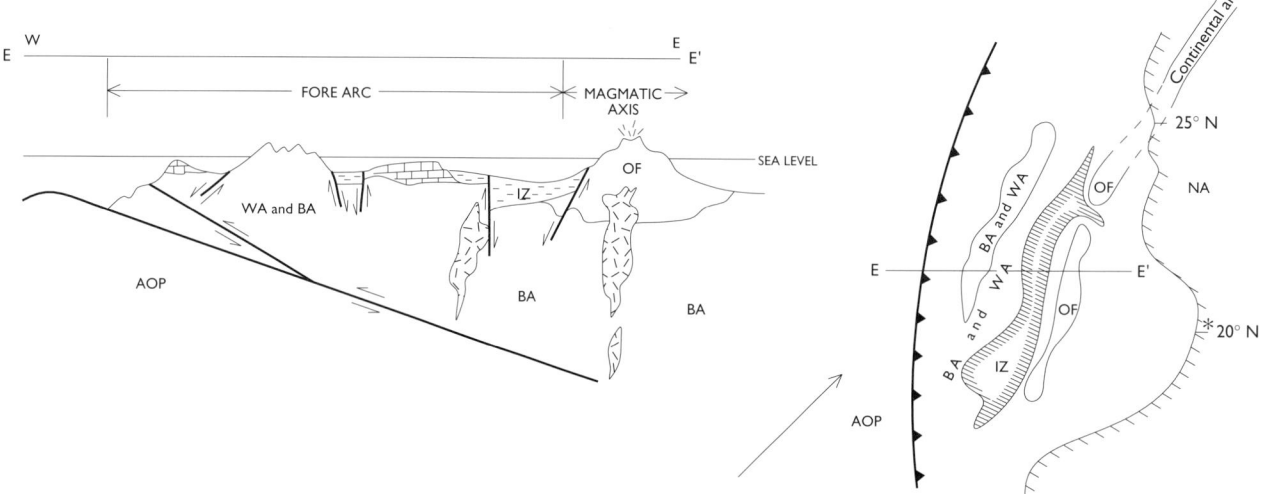

20D. Middle and early Late Triassic (235-225 Ma): This interval was another time of prolific volcanism in the Wallowa Terrane (and Hells Canyon). In the back-arc region I drew in basalt flows, thereby suggesting that other parts of Wrangellia may have been close to the Wallowa Terrane. Basalt flows in the Karmutsen **Formation** of Vancouver Island and in the Nicolai Formation of Alaska (both known to be part of Wrangellia), of the same age as the Wild Sheep Creek Formation in Hells Canyon, have chemical compositions suggesting that they were erupted on an oceanic plateau rather than along the magmatic axis of an island arc (Wallowa Terrane). The back-arc rift indicates that there must have been some kind of fault, or other structure, that separated the various Wrangellia components. Sinistral (left-lateral) faults may have been active at this time within the Wallowa terrane. This is shown in the cross section by a dot and a circle-enclosed cross. The enclosed cross indicates that the portion of the island arc on the right is moving away from the observer. The dot indicates that the portion on the left is moving toward the observer.

F Early and Middle Jurassic (190-155 Ma)

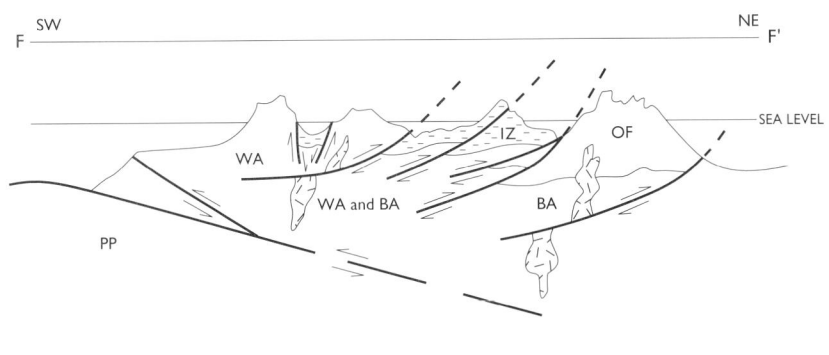
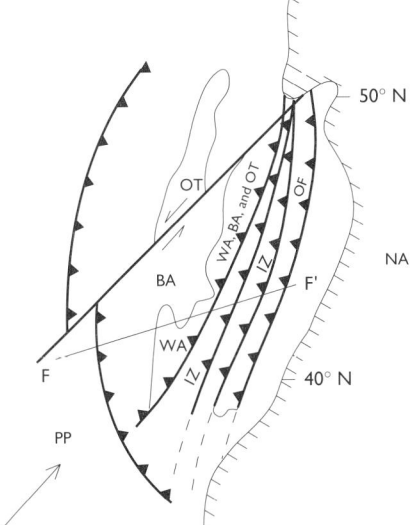

G Late Jurassic to Early Cretaceous (155-115)

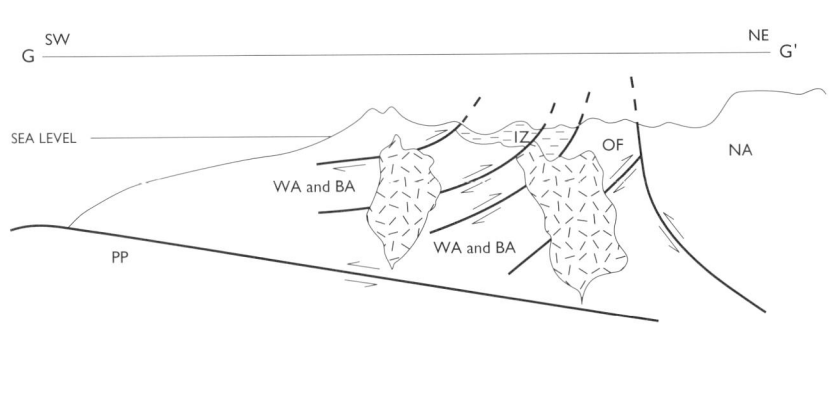
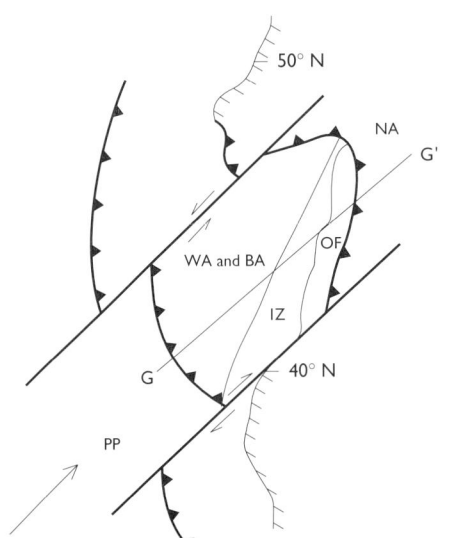

20E. Late Triassic and Early Jurassic (225-190 Ma): I speculate that there was a change in plate vectors in the Late Triassic (early Karnian) such that a new magmatic axis formed (Olds Ferry Terrane) that was active during this time interval. Furthermore, detrital sediments began accumulating in the fore-arc region (Izee Terrane). The Olds Ferry and Izee terranes probably formed on pre-existing rocks of the Baker Terrane. The Wallowa Terrane and large parts of the Baker Terrane also became part of an extensive Olds Ferry fore-arc region. As the outer fore-arc rocks subsided, the Martin Bridge Limestone and equivalent limestone bodies were deposited on those parts of the island arc (Wallowa and Baker terranes) that remained as shallow-water banks and platforms. During this time interval the island arc and continent were both moving northward.

20F. Early and Middle Jurassic (190-155 Ma): This time interval saw the contraction and faulting of the Blue Mountains Island Arc as it approached the North American continent. Many thrust faults formed and parts of all terranes were translated shoreward. No one knows what happened to the ocean floor that must have existed between the island arc and North America. Was it translated northward ahead of the island arc? Was it left behind? Was it subducted eastward beneath North America? Or was it subducted westward beneath the island arc?

20G. Late Jurassic and Early Cretaceous (155-115 Ma): Final movements between the arc terranes and North America occurred around 120 Ma, but this entire time interval was one of collision and welding of the two landmasses. Large batholiths that crystallized in the island arc terranes cut across many of the thrust faults that had formed earlier. By about 120 Ma the island arc had truly become part of North America. At the present time, subduction continues off northern California, Oregon, Washington, and southern British Columbia. The Cascade volcanoes are reminders of what the volcanoes must have looked like on the island arc during its evolution.

After the Late Cretaceous intrusion of the Idaho Batholith, mostly 100 to 70 Ma, the entire region was tectonically lifted into mountains that probably were much higher than those now present. After this uplift, the region was eroded for several tens of millions of years. Beginning in the Eocene and lasting through the Oligocene, volcanoes erupted in central Oregon and some of the flows and ashes covered parts of eastern Oregon. These rocks, however, have not been mapped in the Hells Canyon region, indicating that either they were never there at all or they were eroded off the older rocks before lava flows of the Columbia River Basalt Group were erupted in the middle and late Miocene.

The major Cenozoic event that occurred in the Hells Canyon region was the eruption of the Columbia River Basalt Group, mostly in the 17 to 14 Ma interval. These lava flows extended completely across the present-day Oregon-Idaho boundary and buried all of the pre-Cenozoic rocks beneath thousands of feet of basalt. There is no single accepted reason for the eruption of these flows, but I suspect that the eruptions were associated with the tapping of a deep mantle source in combination with the growth of the Cascade Range and maturing of the Cascadia subduction zone beneath Washington and Oregon. A new regional tectonic pattern, still active today, was imposed on the region after the eruption of the Columbia River Basalt lava flows. The entire region was warped upward and mountain ranges such as the Seven Devils, Wallowa, and Cuddy mountains were differentially uplifted, caused mostly by movement on range-bounding faults. While the region was being raised, rivers and streams tried to reach their base levels of erosion. As a result, the streams have cut spectacular canyons that are deeper now than in the past. A significant event was the emptying of ancient Lake Idaho; its waters raced through Hells Canyon approximately 2 Ma.

Glaciation of mountains in western Wyoming, Idaho, and eastern Oregon influenced the flow volumes and cutting capacities of the streams, particularly as the glaciers melted. I suspect, however, that the fluctuations in water volume did not significantly affect the deepening of Hells Canyon. Glaciers did not reach the bottom of Hells Canyon, although deposits have been mapped in the western Seven Devils Mountains down to about 4,500 feet in elevation (about 3,000 feet above the present level of the river). Glacial deposits can be observed in the upper part of Granite Creek. Wallowa Lake in the northern Wallowa Mountains formed behind a large glacial moraine that now serves as a natural dam (Figure 21).

Figure 21. Wallowa Lake lies within the confines of glacial moraines. Glaciers never reached the bottom of Hells Canyon, although the Seven Devils and Wallowa mountains were glaciated.

CHAPTER 2

Geology of Hells Canyon

GENERAL GEOLOGY

The geology of Hells Canyon is the story of rocks that began their time and space journey somewhere in the ancestral Pacific Ocean (perhaps as many as 300 Ma). The shapes of continents and oceans were much different then; if we were to step back 300 million years in time we would scarcely be able to recognize the surface of our planet. Today, using studies of former landmasses and modern geophysical tools, earth scientists are constructing new maps and configurations of ancient oceans and continents. It is an exciting task, but one that is fraught with oversimplification because of insufficient data.

The Snake River Canyon between Farewell Bend and the mouth of the Grande Ronde River cuts across four terranes of Pennsylvanian (and possibly older) through Late Jurassic age within the complex Blue Mountains Island Arc. These terranes from south to north are the Olds Ferry, Izee, Baker, and Wallowa terranes (Figure 19). Between the Oxbow of the Snake River on the south and the mouth of the Grande Ronde River on the north (Hells Canyon), the only pre-Cenozoic rocks that crop out beneath the flows of the Columbia River Basalt Group are those of the Wallowa Terrane (Figure 22). The geologic history of Hells Canyon (and the Wallowa Terrane) is essentially the evolution of the magmatic axis of an island arc.

Many interpretations of Hells Canyon geology are based on my knowledge of present day island arcs. Although my explanations of Hells Canyon geology may seem somewhat technical, I've tried in places to make the text more understandable to the science layperson. I've left in some details, however, that definitely will be of more interest to professional geologists and geology students.

Prior to my work, the rocks in Hells Canyon were not organized into a comprehensive order. My purpose for mapping these rocks, therefore, was to separate the rocks into mappable units and to organize them in time. This organization of rock units is called a geologic column. In Figure 23 the sequence of rocks in Hells Canyon is placed in a geologic column that shows both the stratigraphic succession and rock ages. The general types of rocks are depicted using symbols. Figure 23 contains both the geologic column for Hells Canyon and a correlation chart. The correlation chart shows the changes that occur within the geologic columns in different parts of Hells Canyon (and the Wallowa Terrane).

Figure 22. Basalt lava flows of the Columbia River Basalt Group overlie the older pre-Cenozoic rocks in Hells Canyon above an **angular unconformity** near Barton Heights.

Figure 23. General geologic column for the rocks in Hells Canyon (bottom) and a correlation of rocks within the Wallowa Terrane (top). Rocks of the Cougar Creek and Oxbow complexes are depicted by an encircled number one in the geologic column. Rocks denoted by x's include not only the Cougar Creek and Oxbow complexes, but also the other Permian and Triassic plutons that are not directly associated with those two plutonic complexes. Triassic (and possibly some Permian) stratified rocks overlie some of the plutonic rocks along an unconformity. Jurassic or Cretaceous plutonic rocks indiscriminately cut across rocks older than Late Jurassic in age.

I divided the stratified, or layered, rocks into rock-stratigraphic units (called formations and groups). Formations are discrete mappable rock units, meaning that geologists can distinguish those units from other units in the region. A group is composed of two or more formations (for example, the Seven Devils Group and the Columbia River Basalt Group). The **intrusive** (or plutonic) rocks are also given names in order to distinguish them from other similar bodies.

The formations of the Seven Devils Group include the Early Permian Windy Ridge and Hunsaker Creek formations and the Middle and Late Triassic Wild Sheep Creek and Doyle Creek formations. Other pre-Cenozoic formations exposed in Hells Canyon are the Late Triassic Martin Bridge Limestone, the Late Triassic and Early Jurassic Hurwal Formation, and the Middle and Late Jurassic Coon Hollow Formation. The Miocene Columbia River Basalt Group is composed of several formations; only two (Imnaha and Grande Ronde) are exposed along the walls of Hells Canyon.

Many of the rock units rest on older rocks along an erosion surface that is called an **unconformity**. An unconformity signifies a gap in the geologic record. The best example of an unconformity is along the base of the first (oldest) flow of the Columbia River Basalt Group where rocks approximately 17 to 14 million years old rest unconformably on other, much older rocks. This unconformity is well exposed along the Oregon side of Hells Canyon, particularly near Wild Sheep and Bull creeks (Figure 22).

The stratified rocks in Hells Canyon are underlain by, and the older stratified rocks are partly intruded by, a complex sequence of metamorphosed igneous (mostly plutonic) rocks. In a few places the plutonic rocks—particularly the dike sequences—form a "basement" to the stratified rocks. These rocks have a wide age range and are well exposed between Suicide Point and upper Pittsburg Landing (Cougar Creek Complex) and near the Oxbow (Oxbow Complex). Other exposures occur between Wolf Creek and the mouth of the Imnaha River and in the southern part of the Seven Devils Mountains. The Jurassic and older rocks are intruded by Late Jurassic or Early Cretaceous plutons.

The Pre-Cenozoic rocks are overlain by some Early Tertiary (older than middle Miocene—probably Eocene) gravels and the Late Tertiary (Miocene) lava flows of the Columbia River Basalt Group. These Cenozoic gravels and lava flows were raised tectonically with the other rocks and, in places like the high Wallowa and Seven Devils mountains, were eroded off the older rocks.

During the late Quaternary, rivers and streams carved deep canyons and left sediments on **alluvial fans** and narrow flood plains. Floods that cascaded through Hells Canyon not only eroded the banks and canyon walls but also left deposits of boulders and gravels. **Landslides**, avalanches, and rock falls have widened the canyons; many of the landslide deposits still remain along the sides of the Snake River and in the upper reaches of some tributary streams.

SNAKE RIVER AND THE CUTTING OF HELLS CANYON

Snake River leaves the relatively flat lava- and sediment-filled Snake River Plain of Idaho at Farewell Bend (Figure 5). From there northward about 160 miles to the mouth of the Grande Ronde River, the river flows in a deep canyon carved in part through the rugged mountains and gentle bordering slopes of the Wallowa Mountains of Oregon and the Seven Devils and Cuddy

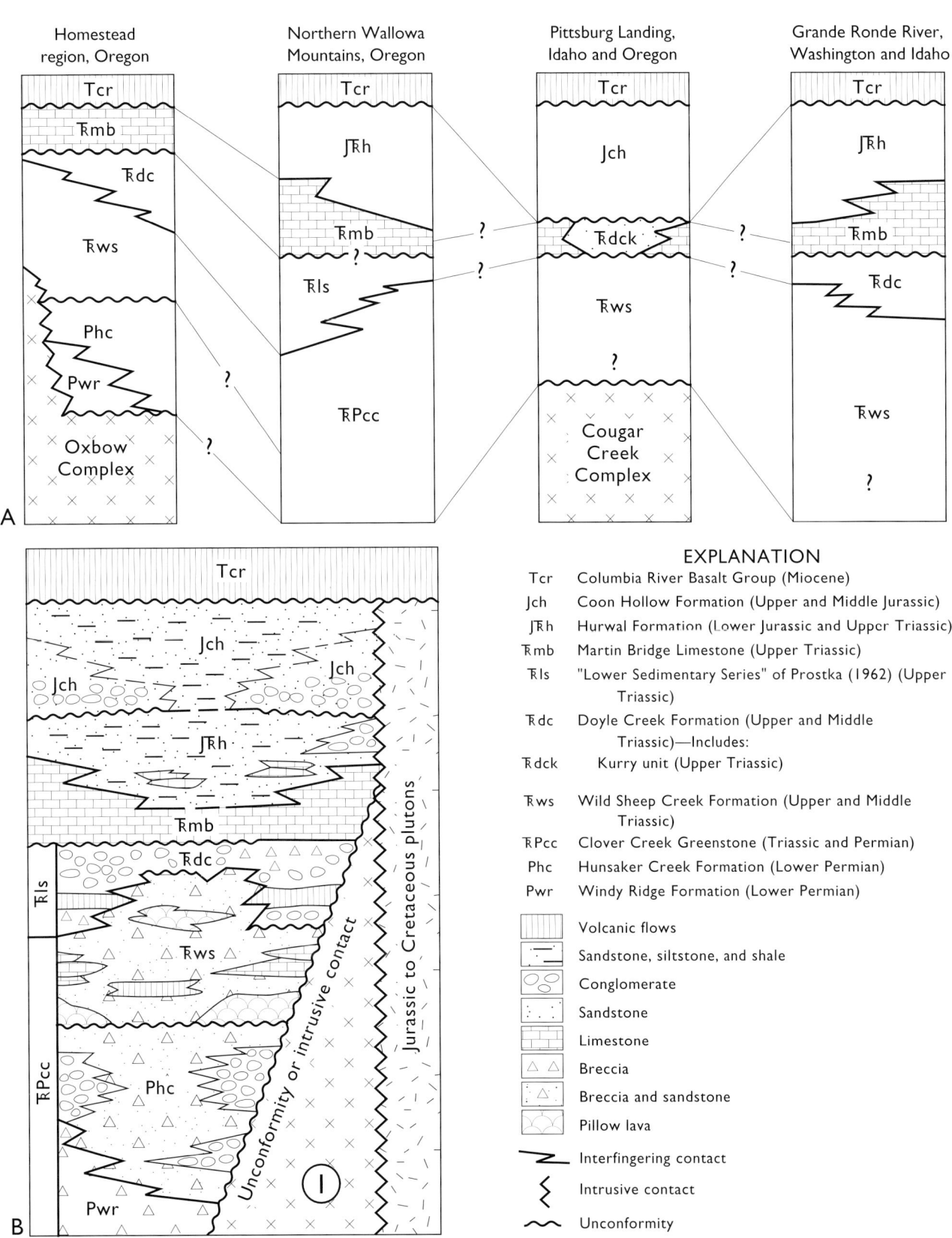

Figure 24. Hells Canyon as seen from a point below Hat Point. This scene has remained relatively unchanged for at least 50,000 years and perhaps even longer.

mountains of Idaho. Beginning near the mouth of the Grande Ronde River, the Snake River cuts through another lava-filled basin and forms a less rugged and much shallower canyon that exists all the way to its confluence with the Columbia River approximately 170 miles farther downstream. The Snake River is impeded by 4 dams between Lewiston, Idaho, and the confluence of Snake and Columbia rivers. Reservoir waters are backed up behind the first of these dams, Lower Granite, all the way to Asotin, Washington.

Between Farewell Bend and Oxbow Dam the river winds through a mountainous area characterized by long narrow ridges, V-shaped tributary canyons, and steep, predominantly soil-covered slopes. For most of this distance, the canyon is only 2,000 to 3,000 feet deep. Below Oxbow Dam, the Snake River plunges into the deep gorge of Hells Canyon. For more than 60 miles, the river is 4,000 to 5,600 feet below the canyon's western rim. To the east, the highest peak of the Seven Devils Mountains towers more than 8,000 feet above the river. Water level of the river is 2,080 feet above sea level at Farewell Bend and water level before impoundment behind Lower Granite Dam at Lewiston, Idaho, was 710 feet. This is a drop of 1,370 feet in 195 miles for an average gradient of about 7 feet per mile, slightly less than the gradient of the Colorado River through the Grand Canyon.

The Snake River has been flowing through Hells Canyon for a long time: probably for no less than 2 million years and possibly for as many as 6 million years. If one were to step back 50,000 years, perhaps even 100,000 years, there would be very little change in Hells Canyon (Figure 24).

Although the canyon is probably a little deeper today, we would still recognize the same tributary streams and even some of the same gravel bars. In places the canyon walls would be much steeper than they are today; in other places the walls would be more gentle.

In a 1954 publication, Wheeler and Cook concluded that the Snake River north of Oxbow was at one time a tributary of the Salmon River. They reasoned that the Snake River eroded its way headward (south) and captured Lake Idaho, which had been drowning southern Idaho for millions of years. The draining of Lake Idaho sent so much water through Hells Canyon that it actually cut a channel deeper than the Salmon River and, consequently, the Salmon River became a tributary of the Snake River. The draining of Lake Idaho probably occurred over thousands of years, but the last remnant of that lake apparently drained in the very latest Pliocene. The youngest mapped unit of lake sediments that had formed in old Lake Idaho, the Glens Ferry Formation, contains **ash** layers that have been radiometrically dated at about 2.5 to 2.0 million years. The lake drained since then.

If Wheeler and Cook are correct, then the Snake River, *as we know it,* began flowing directly to the Columbia River through the canyon about 2 million years ago. The rushing waters caused by the draining of Lake Idaho were important in cutting the deep canyon. However, much more than the water has been responsible for the erosion of this magnificent canyon. The canyon was cut by water as the land was slowly raised. Most of this uplift occurred by movement along faults that caused earthquakes. In fact, if present-day (last 100 years) faulting and earthquakes are the same as those during the past, then they probably have been enough to account for the uplift of the Wallowa and Seven Devils mountains to their present heights in only 2 or 3 million years.

PRE-CENOZOIC ROCKS

The pre-Cenozoic (also called pre-Tertiary) stratified rock succession in Hells Canyon (Figure 23) includes Permian and Triassic lava flows and **volcaniclastic** rocks (Seven Devils Group) that form most outcrops in the canyon, an Upper Triassic platform or shallow-water limestone (Martin Bridge Limestone) with its laterally correlative and overlapping marine sandstones and argillites (Hurwal Formation), and Middle Jurassic **fluvial** rocks that are overlain by Upper Jurassic marine sandstones and siltstones (Coon Hollow Formation).

The pre-Cenozoic stratified rocks were intruded throughout their histories with plutonic phases (predominant times of crystallization) in the Permian (270 to 255 Ma), Triassic (245 to 230 Ma), and Jurassic-Cretaceous (140 to 115 Ma) intervals. Along the border zone with the ancient North American continent in western Idaho (Salmon River suture), some plutons are 110 to 95 million years old. East of the border zone in west central Idaho, rocks of the Idaho Batholith crystallized within the 100 to 70 Ma interval, mostly 85 to 70 Ma, and some plutons crystallized as late as the Eocene.

Except for the younger plutonic rocks (those of Late Jurassic and Cretaceous ages) almost all of the pre-Cenozoic rocks in Hells Canyon have been variably metamorphosed. **Metamorphism** (caused by an increase in pressure and temperature during earth movements and the crystallization of **magmas** that formed the plutonic bodies) rearranged the ions in minerals to make new minerals that were stable under the new pressure and temperature regimes. During metamorphism, sandstones can become **gneisses** and mudstones can become **schists**. In the older rocks of Hells Canyon, most minerals were changed during metamorphism. The textures remained relatively unchanged, except in the greatly deformed rocks of the Cougar Creek and Oxbow complexes.

In Figure 23, notice how the rock sequences change from south to north in Hells Canyon (from Homestead through Pittsburg Landing and then to the mouth of the Grande Ronde River). The rock sequences also change from east to west between Hells Canyon and the Wallowa Mountains. Names other than those of the Seven Devils Group (Clover Creek Greenstone and "Lower Sedimentary Series") were previously given to stratified Permian and Triassic rocks in the Wallowa Mountains. Rock formations in the Wallowa Mountains are still recognized by those earlier assigned names.

Stratified Rocks

The geologic history of Hells Canyon is tied closely to an understanding of the stratified (layered) rocks. Their fossils and sedimentary characteristics tell geologists about source areas, transporting mechanisms, and environments of deposition.

The units of Pre-Cenozoic stratified rocks (lava flows and sedimentary rocks) in Hells Canyon were erupted and deposited within relatively short periods of time during the Permian (270 to 260 Ma), Triassic (235 to 220 Ma), and Jurassic (180 to 175 and 165 to 160 Ma) periods. The short time intervals suggest that volcanic activity along the magmatic axis of the Blue Mountains Island Arc was relatively short-lived. Such volcanic activity is comparable to the volcanic activity of modern island arcs where relatively short episodes of volcanism have been interrupted by long periods of quiescence.

The Permian rocks record volcanic activity on ancestral Pacific islands during early growth phases of the island arc. The Triassic rocks record prolific volcanism along the volcanic axis of the island arc, followed by erosion of the islands, subsidence, and overlapping non-volcanic sedimentation. The Jurassic rocks record **subaerial** volcanic activity, erosion, and sedimentation, followed by subsidence and consequent **transgression** of the ocean and, finally, the deposition of sediments from deep water **turbidity currents**.

Permian strata are divided into the Windy Ridge and Hunsaker Creek formations. Both of these rock units are well exposed in the southern part of Hells Canyon near Oxbow, Oregon. The Windy Ridge Formation (Pwr on accompanying geologic maps) is composed of fragmental debris, consisting mostly of pyroclastic **breccia** and **tuff**. Rare lava flows are intermixed with the fragmental rocks; basalt and **diabase** dikes are common. Minerals within rocks of this formation are mostly **quartz**, albite, magnetite, **epidote**, and **chlorite**. The original stratified rocks were potassium-poor **rhyolite** tuffs and flows, but because of the greenschist-**facies** metamorphism, they are now referred to as **quartz keratophyre** tuffs and flows. The original glass groundmasses of flows and matrices of tuffs are replaced by a fine mosaic of intergrown quartz and albite. Large quartz crystals have a blue tint, indicating that they contain a small amount of included water. Although no fossils have been recovered from this unit, I presume that the Windy Ridge Formation is of Early Permian age because the tuffaceous rocks of that formation interfinger with the lower part of the Hunsaker Creek Formation, which contains fossils of Early Permian age.

The Early Permian Hunsaker Creek Formation (Phc on accompanying geologic maps) is a diverse collection of the following rocks: pyroclastic breccia and tuff (Figure 25), **conglomerate**, sandstone, breccia, and siltstone with abundant diabase, basalt, **andesite**, and rhyolite dikes and **sills**. Rare lava flows of basalt, andesite, and rhyolite compositions occur in the strata. **Lithofacies** changes occur over very short distances indicating the heterogeneity of depositional depths and slopes as one might expect in island settings. Fossils include productid and spirifer brachiopods and some clams. It is significant that Permian strata in the Baker Terrane contain fusilinids and radiolarians, suggesting that the sediments were deposited in an oceanic setting rather than in an insular setting. The absence of fusilinids in the Wallowa Terrane may also have paleoclimate implications (colder water).

Quartz-rich compositions of the sedimentary rocks indicate an abundance of rhyolitic source rocks, although andesitic and basaltic sediments are common. Metamorphism to the **greenschist facies** changed the rocks to quartz keratophyre (meta-rhyolite), keratophyre (meta-andesite) flows and tuffs, meta-basalt lavas and dikes, and sediments that consist mostly of quartz and albite, but with minor amounts of epidote, calcite, and chlorite. Outstanding outcrops occur in the southern part of the Hells Canyon region between Oxbow and the Kleinschmidt Grade on both sides of the canyon. The rocks were originally lava flows, tuffs, and sediments that had been both erupted and eroded from volcanic islands. Associated dikes and sills of the Permian intrusives (Pi on included maps) are characteristic of the small intrusive bodies found in the Permian throughout the mapped area. The volcanoes were highly explosive, similar to those in modern island arcs such as the Aleutians.

An estimated thickness range of the Windy Ridge Formation is between 1,000 and 1,500 feet and that of the Hunsaker Creek Formation is between 7,500 and 10,000 feet.

Triassic stratified rocks rest unconformably on Permian rocks. In the lower (older) parts of the Triassic section, the rocks are predominantly volcanic (lava flows, dikes and sills, pyroclastic breccia and tuff) and **epiclastic** conglomerate, breccia, sandstone, and siltstone. These rocks are assigned to the Wild Sheep Creek and Doyle Creek formations. In the upper (younger) parts of the Triassic section, the rocks are almost all sedimentary. These sedimentary rocks are assigned to the Martin Bridge Limestone and Hurwal Formation. The Martin Bridge Limestone consists predominantly of limestone, and less commonly of breccia, sandstone, and siltstone. Rocks in the Hurwal Formation consist primarily of sandstone and siltstone with less common conglomerate, breccia, and limestone.

The oldest Triassic stratified rock unit is the Wild Sheep Creek Formation, named for outstanding outcrops in the Wild Sheep Creek area. In general, the steepest walls of Hells Canyon are composed of rocks of the Wild Sheep Creek Formation. In contrast to the Permian quartz-rich rocks, rocks in the Wild Sheep Creek Formation are more **mafic**. (They contain more magnesium and iron.) The Wild Sheep Creek Formation consists of dark green, black, and maroon flow rocks, pyroclastic rocks containing fragments of basalt, **basaltic andesite**, and andesite, and sedimentary rocks derived from the erosion of lava flows and pyroclastic rocks. Pillowed lava flows and pillow breccias are common in parts of the canyon and in the Seven Devils Mountains. Breccia, however, is by far the dominant rock type in the Wild Sheep Creek Formation; most of these coarse-grained sediments apparently cascaded down steep submarine slopes as **debris flows**. In places, thick

Figure 25. Tuff beds in the Permian Hunsaker Creek Formation.

sequences of sandstone and siltstone in **graded beds** were deposited from turbidity currents. Outcrops of limestone commonly occur in the Wild Sheep Creek Formation north of the Imnaha River-Snake River confluence. Some limestone bodies are slide blocks that slid along submarine slopes from shallow to deep water. Other limestone bodies interfinger with volcanic breccia and sandstone. The volcanic rocks that have been metamorphosed to the greenschist facies contain albite, epidote, magnetite, chlorite, and calcite as common minerals. Quartz, white mica, **hematite**, and pyrite are rare. Relict (crystallized from magma or lava) labradorite and clino**pyroxene** are abundant in some rocks from outcrops where metamorphism was less intense. The maximum thickness of the formation may be as much as 10,000 feet, and the age is Middle and Late Triassic (late Ladinian and early Karnian).

APRIL, 1982. I WAS RELIEVED *from the midnight to noon watch and wandered onto the deck of the research vessel S. P. LEE. We were working in the southwest Pacific Ocean near the islands of Tonga and, at the time, had airguns and a long hydrophone streamer over the fantail as we seismically probed for basins that might contain oil and gas. Ata, the southernmost island of the Tonga Island Arc, loomed on the horizon. Ata is an active volcano, known to have erupted in historic time.*

I wanted samples of lava flows from Ata Island to compare with those we had dredged from the western scarp of the Tonga Ridge. I asked the first mate, Dave, and the third engineer, Bill, to accompany me to the island in a small Zodiac boat. We took survival suits, sample bags, food and water, sleeping bags, a first aid kit, extra gasoline, and a hand-held radio sealed in a water-tight bag. As I put on my work boots and ball cap, Dave lowered the Zodiac over the side with a crane. Bill grabbed the gasoline, food, and water. We planned an overnight stay, catching the ship as it returned on the next seismic line-sometime after lunch the next day.

At the closest point to Ata, about two miles distance, the ship slowed to four knots. We jumped into the Zodiac, and I started the outboard motor just as we reached the water. Bill unhooked the cables and we headed for the uninhabited island; the S. P. LEE continued on its preset trackline.

The mid-day sun reflected off pale blue waters. The Zodiac slowly approached the island as the ship steamed over the horizon and out of sight. The island is small: less than a mile in diameter and three miles in circumference. We slowly circled the island just outside the surf zone, finding no obvious place to land. In most places the waves crashed against steep barren cliffs, but finally a small pocket beach appeared along the west side of the island. We headed for it.

The waves were much larger than we had thought; they curled sharply near the proposed landing spot. Dave was handling the boat as we approached the surf, so I instructed him to ride a wave as far as possible and to cut the motor just before we hit the beach. Unfortunately, he panicked at the top of a large wave and cut the motor too soon. The boat lost its momentum and the power of the wave catapulted us into the surf. We were swept under the water. The boat came down on top of Dave, slamming him into the sand and water. When my feet felt the sand I bobbed upright and saw Dave floating back out to sea with the boat. I yelled to Bill and told him to grab the boat, then waded out and nabbed Dave by the shirt collar. He was dazed but conscious. Blood ran across his face from a jagged scalp wound. I pulled him through the surf and onto the beach while Bill collected some of the flotsam that had fallen out of the boat. I sat Dave on a flat rock, opened the first aid kit, and dressed his wound. I wrapped a large blue bandage around his head, tore the ends and knotted them. He was still dazed and wanted to sleep, but I convinced him that he might have a concussion and had to stay awake.

The tide came in and soon the beach was covered by water, leaving a small rock terrace about twenty feet above the beach. Bill and I carried the boat and gear to the top. The terrace was no more than fifty feet long and fifteen feet wide. When large waves broke, the water sprayed us on the terrace and kept us uncomfortable and wet. I tried to call the ship, but the volcano was between us and the ship so the ship didn't answer. The sun was setting. We knew the night would be a long one. Our sleeping bags were wet and we had only six oranges and a jar of peanut butter among us. A full moon rose as the last flash of sunlight graced the western horizon, turning it deep crimson. For dinner we ate an orange apiece, one-third of the peanut butter, and drank some water from the canteens. I stayed awake all night, checking on Dave hourly and watching the moon drift across the sky. I worried that we wouldn't be able to start the outboard motor for the journey to the ship and that we would once again capsize in the surf.

The next morning we ate the last oranges and finished the peanut butter. I poured more antiseptic on Dave's scalp wound and changed the bandage. He was still somewhat confused. As Bill cleaned the spark plug and dried out the boat, I trekked along the cliffs to collect rocks.

The layers of lava flows and breccias looked very similar to some of the stratigraphic sequences in the Triassic Wild Sheep Creek Formation of Hells Canyon. Some lava flows were pillowed, others were not. Dikes and sills cut the stratified rocks. At the time I thought that I might be able to compare the chemical composition of these lava flows with those in the Wild Sheep Creek Formation of Hells Canyon. Later on, I found that the chemistries are very similar. Again, the "present is the key to the past." Lavas in the Wild Sheep Creek Formation erupted on islands and seamounts along an island arc that was similar to the Tonga Island Arc.

The morning tide was ebbing so I jumped and crawled along the cliffs without getting drenched. Around lunch time I returned to find Dave and Bill anxious to go. By that time the tide was out. They had inflated the boat and Bill was pretty sure that the motor would start. However, we still had a problem: the waves were large and breaking hard against the small pocket beach. We carried the boat down to the beach where I sat on a rock to count the waves, hoping there was a periodicity of the small and large waves. Sure enough, every seventh and eighth waves were the largest. We prepared to launch the boat while I counted the waves, "One, two, three, four, five, six, seven (large), eight (large)." I shouted, "Okay, let's go." We carried the boat into the water. I jumped into the boat. Then Bill and Dave pushed the boat farther out while I pulled the starter rope. The motor started on the first pull. They vaulted into the boat and we planed gracefully through the surf to reach the open ocean.

Data from the Ata Island rocks have resulted in several publications. Members of the scientific community now know the age, petrology, and geologic history of Ata Island, one of the young volcanoes of the Tonga Island Arc. And scientists also know that Ata Island rocks are very similar to those in the Wild Sheep Creek Formation of Hells Canyon.

Figure 26. Boulder of pyroclastic breccia in the Triassic Doyle Creek Formation above Eagle Bar near the trail to the Red Ledge Mine, southern Hells Canyon.

All the rocks of the Doyle Creek Formation are maroon, red, and green including epiclastic conglomerate, breccia, sandstone, and siltstone, pyroclastic breccia and tuff (Figure 26). The formation has only minor amounts of lava flow rocks. Many clastic rocks in the Doyle Creek Formation are subaerial and shallow-water marine equivalents to submarine rocks in the upper part of the Wild Sheep Creek Formation; they are lithofacies. In other places, the Doyle Creek Formation consists mostly of maroon pyroclastic deposits; whereas in others, maroon conglomerate and sandstone are dominant. At Pittsburg Landing, the Kurry unit of the Doyle Creek Formation consists of limey and ash-rich sandstone and siltstone. The thickness of the Doyle Creek Formation has a wide range, from zero to possibly as much as 1,500 feet, and its age is Middle (?) and Late (early Karnian) Triassic.

The Martin Bridge Limestone rests unconformably on the older Triassic rocks. It is of Late Triassic (late Karnian and early Norian) age in the region; only early Norian fossils have been identified within the Martin Bridge Limestone in Hells Canyon. Although it is named "limestone," rocks of this formation in the Wallowa Mountains and near the mouth of the Grande Ronde River also consist of noncalcareous sandstone, breccia, and siltstone. Furthermore, the Martin Bridge Limestone intertongues with, and is in part equivalent to, the lower part of the Hurwal Formation. The Martin Bridge Limestone in Hells Canyon at Kinney Creek is a shallow-water platform limestone; near the mouth of the Grande Ronde River, the formation has more clastic material than at Kinney Creek and intertongues with, and is overlain by, the Hurwal Formation. In the southeastern Wallowa Mountains near Summit Point a spectacular coral reef occurs in the Martin Bridge Limestone. The Martin Bridge Limestone is more than 1,500 feet thick at Kinney Creek.

The Hurwal Formation is exposed in Hells Canyon only on the Idaho side of the Snake River near the mouth of the Grande Ronde River. No fossils have been found in outcrops there, but elsewhere, in the Wallowa Mountains, the Hurwal Formation contains abundant fossils (including **ammonites**) and is Late Triassic and Early Jurassic in age. The Hurwal Formation consists mostly of sandstone, siltstone, and breccia. In the northern Wallowa Mountains the Hurwal Formation contains large olistoliths (slide blocks) of limestone. These limestone blocks broke off shallow-water platforms that surrounded the islands and slid into deeper water. The limestone was forming in shallow water while adjacent deep basins were being filled with siliciclastic (noncarbonate) detritus eroded from the islands of the volcanically quiescent and fragmented terranes. Similar large blocks, both of limestone and basalt, have been mapped near the Hawaiian islands chain where parts of shallow-water reefs, other limestone deposits that formed on platforms surrounding the islands, and thick basalt sequences broke off and cascaded downslope onto the deep ocean floor.

The Jurassic Coon Hollow Formation is well exposed at Pittsburg Landing and along Hells Canyon near the Washington-Oregon border. Rocks at Pittsburg Landing show a change from bottom to top: from tuffaceous rocks to subaerially deposited conglomerate and sandstone, then to marine sandstone, and finally to **turbidite** beds of calcareous sandstone and siltstone. Farther north near the Washington-Oregon border, the rocks are conglomerate near the base and turbidite sandstone and siltstone at higher stratigraphic elevations. The subaerial older parts of the Coon Hollow Formation are Middle Jurassic (Bajocian) in age, whereas the younger marine parts are Late Jurassic (Callovian and Oxfordian) in age.

Intrusive Rocks

Miles beneath the volcanic and sedimentary carapaces of the Wallowa Terrane, intrusive (plutonic) rocks crystallized from magmas that had formed in the crust and mantle above a subduction zone. Some of the magmas also fed the lava flows that erupted onto the surface of the sea floor and islands. In places, volcanoes grew above the magma conduits. In other places, the lava flows erupted from fissures. Many of these intrusive rocks have the same chemical composition as the lava flows, but the mineral crystals had a longer time to crystallize and, thereby, are larger (coarser grained) than those in the lava flows.

Intrusive rocks in Hells Canyon exhibit a wide range of ages, compositions, and textures. All of the pre-Cenozoic intrusive rocks, however, are similar to those described from Cenozoic island arcs that border the Pacific Ocean. Most of the larger plutonic rock bodies consist of gabbro, **norite**, diorite, quartz diorite, and **trondhjemite**. Granodiorite and granite plutons are rare. Intrusive rocks of Permian and Triassic ages are remarkably potas-

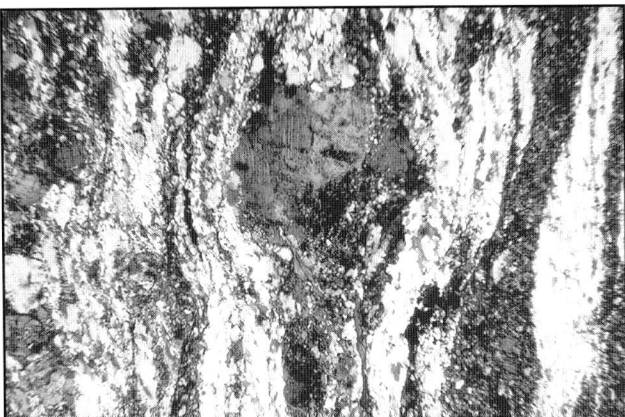

sium poor. Dikes and sills have compositions that mimic the volcanic flows—every composition from basalt (including diabase) through andesite, **dacite**, and rhyolite.

Large bodies of aligned dikes occur mainly near the Oxbow of the Snake River (Oxbow Complex) and along several miles of the canyon between Temperance Creek and Pittsburg Landing (Cougar Creek Complex). The dikes range in thickness from a few inches to tens of feet and have compositions ranging from gabbro (and basalt) to quartz diorite (and rhyolite). The rocks have been variably metamorphosed to the greenschist and **amphibolite facies** and, in places, the rocks have been broken, sheared, and recrystallized to form **cataclasite**, **mylonite**, and **gneissic mylonite** (Figures 27 and 28). The Cougar Creek Complex contains rocks that have yielded the oldest radiometric age in Hells Canyon (about 300 million years). Dikes in the Cougar Creek Complex are cut by small, and essentially undeformed, quartz diorite **stocks**; two of these stocks have radiometric ages, respectively, of 256 and 246 million years.

Only a few intrusive rock bodies in Hells Canyon have been radiometrically dated. Many of those that crystallized in the Permian and Triassic are similar in age to the stratified rocks of the Seven Devils Group. I presume that these plutons are the crystallized **magma chambers** of the lava flows and tuffs of the Seven Devils Group.

Jurassic to Cretaceous plutons, such as those in the Wallowa Batholith, are uncommon in Hells Canyon and occur within relatively small geographic areas. The best exposures are in the upper part of Deep Creek and in Granite and Little Granite creeks. The youngest Jurassic to Cretaceous pluton, exposed along Granite Creek, has been radiometrically dated and is approximately 115 million years old.

CENOZOIC ROCKS

Tertiary Gravels (pre-Columbia River Basalt Group)

Unconsolidated boulders, cobbles, pebbles, and sands commonly crop out beneath flows of the Columbia River Basalt Group. They comprise a wide assortment of rock types and most were derived from erosion of Hells Canyon rocks and of the Idaho and Wallowa batholiths. However, quartzite is a

Figure 27. This mylonite in the Oxbow Complex was originally massive trondhjemite. The trondhjemite was shattered by tectonic forces related to faulting and recrystallization occurred with the mineral folia forming nearly perpendicular to the maximum stress direction. Pencil is shown for scale.

Figure 28. Photograph of a thin section (photomicrograph) of mylonite from the Oxbow Complex. The thin section was photographed in polarized light. Thin sections of rocks (about .035 mm thick) are used by geologists to study mineral compositions, textures and structures under fairly high magnification. The field of view is about 3.0 mm long. The rock consists mostly of feldspar, quartz, and hornblende (darker shade). A feldspar crystal (center) somehow resisted the grinding forces.

Figure 29. Dozens of lava flows (Columbia River Basalt Group) form rugged outcrops in Oregon, north of Hat Point *(Color section, page 92)*.

Figure 30. Distribution of the Columbia River Basalt Group and associated volcanic rocks (figure modified from Hooper and Swanson, 1990, with permission from Peter Hooper). The shaded area includes rocks of the major formations in the Columbia River Basalt Group: Imnaha Basalt, Grande Ronde Basalt, Wanapum Basalt, and Saddle Mountains Basalt, plus the related volcanic rocks that include the Picture Gorge Basalt, basalt of Powder River, basalt of Prineville, and basalt of Weiser. Asterisks mark active Cascade volcanoes. KBML is the Klamath-Blue Mountains Lineament, which forms the western boundary of the present-day Blue Mountains in northeastern Oregon.

ubiquitous component in all gravels and is the sole component of some. Outcrops of these Tertiary gravels, tens of feet thick, occur on the eroded and beveled surface of Martin Bridge Limestone near the mouth of the Grande Ronde River in northern Hells Canyon. Throughout the Hells Canyon region, and even in places on the Wallowa Batholith, boulders are scattered on eroded surfaces or occur in thin-bedded deposits.

Quartzite boulders and cobbles are scattered like watermelons along the eroded surface of Martin Bridge Limestone between McGraw and Spring creeks in southern Hells Canyon. The largest boulders are about two feet in diameter; common percussion marks on the boulders indicate that the boulders struck each other forcefully during transport in a raging stream. Quartzite strata do not occur in the Wallowa terrane; it is probable that the quartzite boulders were transported by streams from outcrops in north-central Idaho.

These Tertiary gravels were eroded from mountainous terrains in Idaho and eastern Oregon after the Blue Mountains Island Arc had been zippered or accreted to the North American continent. The gravels could be of any age from Late Cretaceous to early Miocene. Based, however, on geologic interpretations of similar gravels elsewhere in Idaho, Oregon, and Montana, I presume that the gravels were deposited mostly during the Eocene Epoch.

Columbia River Basalt Group

Lava flows of the Columbia River Basalt Group form the highest levels of Hells Canyon, and are particularly evident on the Oregon side (Figure 29). The lava flows in Hells Canyon represent only a small section of the extensive Columbia River Basalt Group. The Columbia River Basalt Group has an age range of about 17 to 6 Ma, but flows in Hells Canyon erupted mostly in the 17 to 14 Ma interval. North of the confluence of the Snake and Grande

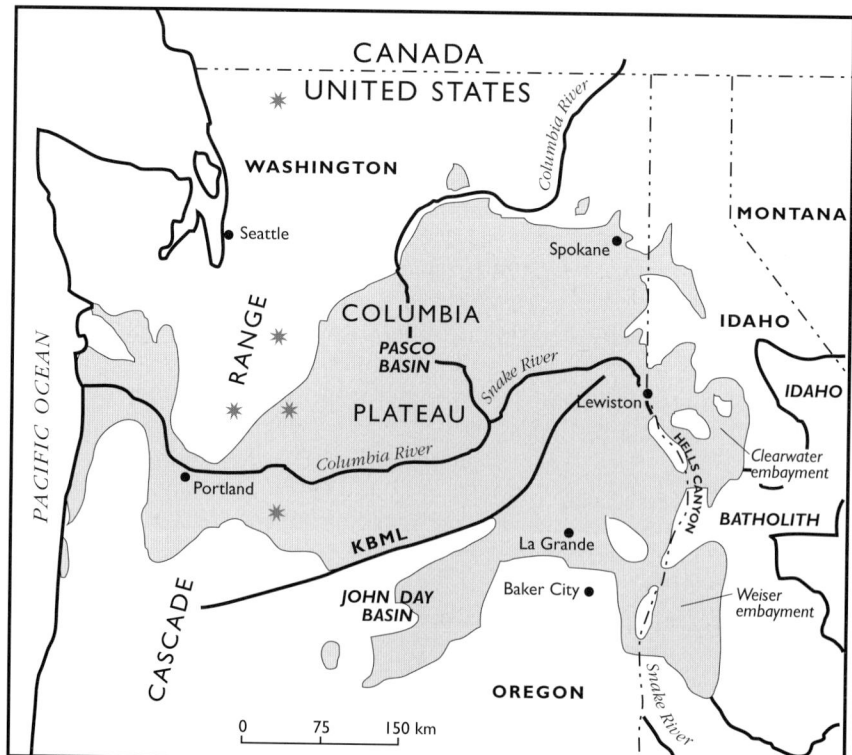

Ronde rivers, many of the lava flows are younger than those that line the sides of Hells Canyon farther south. Some of these younger lava flows were erupted 12 to 10 Ma. Isolated lava flow sequences near Lewiston, Idaho, are as young as 6 million years. Long intervals of time often separated successive lava flows.

Outcrops of the Columbia River Basalt Group provide spectacular geologic scenery in the Pacific Northwest. Flows of basalt embellish deep canyon walls throughout large parts of the region, particularly in central and eastern parts of Washington and Oregon. The area presently covered by flows of the Columbia River Basalt Group is about 164,000 square miles (Figure 30). Some individual flows, as thick as 300 feet, cover thousands of square miles.

Lava flows of the Columbia River Basalt Group in Hells Canyon were erupted onto the earth's surface over a relatively short time period. If we assume that there were a maximum of 100 flows, and that these flows were extruded over a period of only 2 million years, then on an average one flow was erupted every 20,000 years. I doubt if the eruption cycle of flows was that regular. No matter what their cyclicity, however, we can conclude that there were some very long time intervals between eruptions. During those quiescent intervals, soils developed, flowers and trees grew, animals and insects walked, crawled, and flew over lush prairies of grasses and vast forests of trees. Generations of plants and animals lived, evolved, and died between eruptions. The red colors that can be seen between lava flows are soil zones that had formed on top the preceding lava flow. Heat from the succeeding lava flow baked the soil and oxidized the iron, thereby making the rusty-red color.

The land surface in the Hells Canyon region was subdued when the first flow of the Columbia River Basalt Group was erupted. Whereas the maximum relief in the region (highest peak in the Wallowa Mountains to river level at the mouth of the Grande Ronde River) is presently more than 9,000 feet, the maximum relief at the time of the first lava flow was probably no more than 3,000 feet. During the last approximately 6 million years, tectonic uplift has formed (and is still forming) the rugged mountainous relief we see today in the Hells Canyon and Wallowa Mountains regions.

The geology of the Columbia River Basalt Group is well known. It is divided into four major formations (Figure 31). Several formal members and many named lava flows further subdivide this vast pile of lava. The Imnaha and Grande Ronde formations are dominant in the southern and central parts of Hells Canyon. Most basalt flows were erupted from low-relief (probably linear) vents. Conduits for the magma that erupted as lava flows are preserved as dikes (feeder dikes); some of the dikes occur as dark-colored stripes and stringers that stand out in stark contrast to the lighter colored pre-Cenozoic rocks; this is particularly evident in the northern Wallowa Mountains where black and dark brown dikes cut the white and gray granitic and limestone rocks. Extensive dikes also can be observed in the eastern part of the Grande Ronde River Canyon.

LATE QUATERNARY DEPOSITS

A long period of erosion occurred between the eruption of the youngest basalt flows in the late Miocene and the late Quaternary. Some major gravel and boulder beds occur in the Lewiston-Clarkston-Asotin region that both

Columbia River Basalt Group	Age (Ma)
Saddle Mountains Basalt	6.0 — 13.5 —
Wanapum Basalt	
Hiatus	14.5
Grande Ronde Basalt	— 16.5 —
Imnaha Basalt	— 17.5 —

Figure 31. Formations of the Columbia River Basalt Group (modified from Hooper and Swanson, 1990). Parts of all formations are exposed in eastern Oregon, southeastern Washington, and western Idaho near Hells Canyon with the youngest flows in the upper elevations between the Grande Ronde River and Clarkston, Washington, and in the Lewiston-Clarkston area. Formation thicknesses are not shown in this diagram.

interfinger with and overlie some of the youngest flows of the Columbia River Basalt Group indicating that they are late Miocene to late Quaternary in age. The late Quaternary deposits, however, are especially important contributors to the geologic story of Hells Canyon and give some predictions about future geologic events. I focus on them here. Landslides, floods, and earthquakes were common in the past and will occur in the future.

Landslide and Slump Deposits

Landslide and **slump** deposits form several topographic features in the Hells Canyon region. A spectacular landslide created part of Big Bar in the Snake River Canyon approximately seven miles south of Hells Canyon Dam.

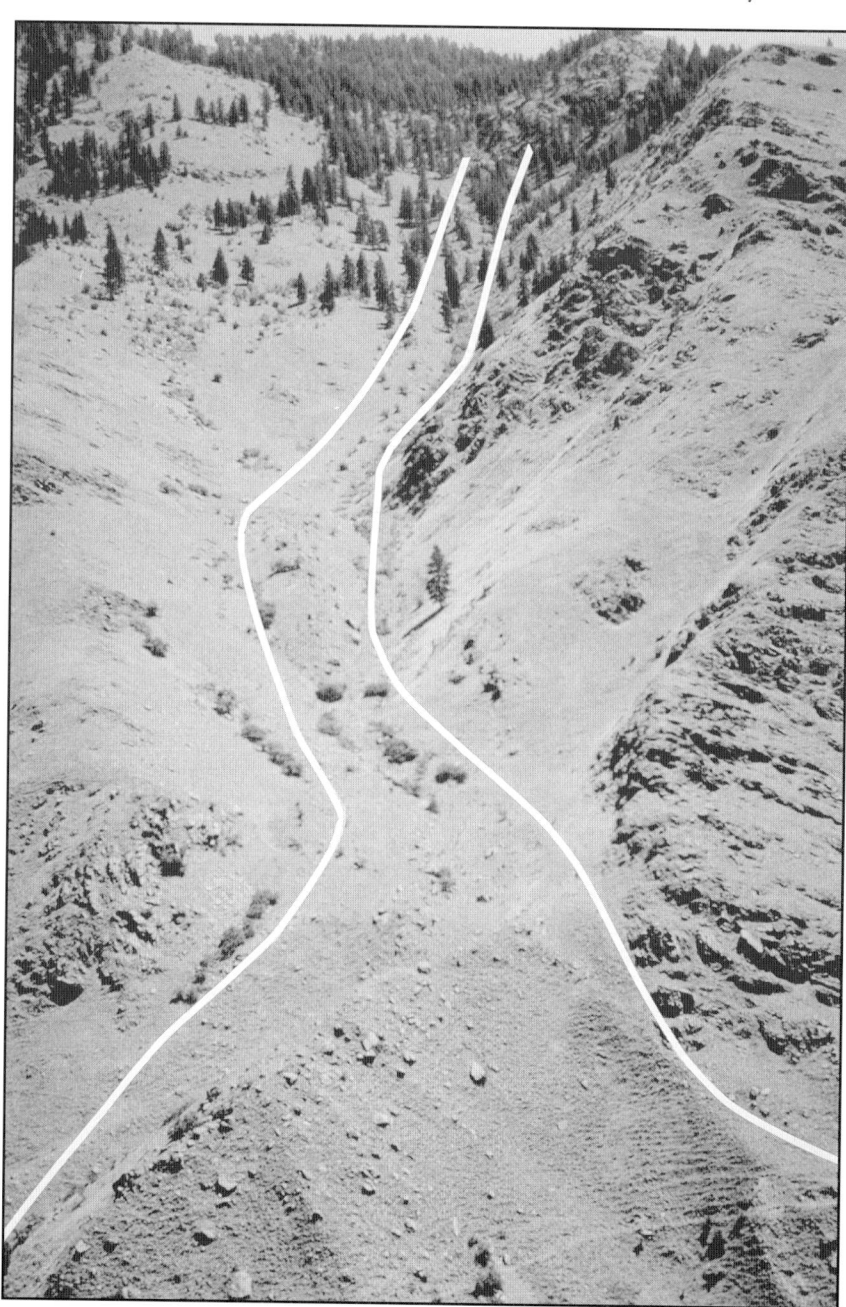

Figure 32. White lines define the chute that guided landslide debris from high above the Snake River in Idaho to form a dam nearly 400 feet high at the mouth of Rush Creek. The Snake River is near the base of the photograph.

Other landslide (and slump) deposits can be observed along the Snake River (from south to north) at its confluence with Copper Creek, near Bernard Creek, at Rush Creek, between Marks and Waterspout creeks, and at Johnson Bar. Portions of the hillside above Johnson Bar are gashed by small slump scarps that forecast future landslides. Sections of that hillside may begin to move again after heavy rainfall saturates the strata, particularly if an earthquake were to simultaneously shake the canyon slopes (Figure 32).

Landslides should be expected at any time in some parts of both Hells Canyon and the tributary canyons of the Snake River. Many of the canyon walls are precipitous, rocks are crumbly and severely weathered, and relatively large earthquakes (as strong as Richter **magnitude** 5, and possibly as strong as magnitude 6) have apparently occurred in the past and should be expected in the future. In certain places, the strata dip steeply toward the river, particularly on the Idaho side in the deeper parts of Hells Canyon near Hells Canyon Dam. Over saturation by water and shaking by earthquakes could trigger a significant landslide there. It is impossible to predict where or when future landslides will occur. Areas that I think are most susceptible, however, are those where the canyon walls are very steep: (1) near Eagle Bar, in the Hells Canyon Dam area (specifically on the Idaho side about a mile below the dam), (2) between Bernard Creek and Johnson Bar on the Idaho side of the river, and (3) at the Eagle's Nest outcrop in Oregon.

A spectacular landslide feature is now partially submerged in the Hells Canyon Dam reservoir at Big Bar (the area in Idaho near the confluence of Allison Creek and the Snake River). A landslide had cascaded into the river after a slope failure in Oregon. Bonneville Flood waters deluged the landslide debris and, due to the resultant loss of energy caused by increased friction and lower velocity caused by the wider canyon floor, abundant debris was deposited that previously had been contained within the flood waters. Now the rounded apex of the landslide feature barely pokes through the water surface as an island in the middle of the reservoir.

River Terraces

River **terraces** are the products of several processes. The power of river flow alone can erode solid rock and create flat rock terraces. High velocity river currents, such as those associated with floods, often flow over alluvial fans and landslide debris, thereby planing off the sediment and forming a relatively flat terrace. If there is abundant sediment in the water and the water velocity is significantly slowed by a temporary dam or by a constriction of the river channel, then sediment will be deposited on the floor of the affected temporary lake. Subsequently, a new channel will be cut through the lake floor, leaving terraces behind.

River terraces, some rising more than 100 feet above the present river level, are well exposed in several parts of the canyon (Figure 33). The terraces have not been studied in much detail, but it is apparent that the Bonneville flood had a strong effect on these high terrace configurations. There is no doubt, however, that other major floods have also formed some high terraces. Outstanding examples of high terraces can be observed in the area between Rush Creek and Johnson Bar, between Quartz and Temperance creeks, at High Bar, and at Pittsburg Landing. The lower terraces that parallel the river banks (ten to fifteen feet above water level) were formed by floodwaters in the past century, probably as the result of extremely high water during spring thaws.

38 ISLANDS AND RAPIDS

Figure 33. Terraces that line the Oregon bank of the Snake River below Rush Creek formed during catastrophic floods. The highest terrace probably formed during the Bonneville Flood.

Figure 34. Alluvial fan at the mouth of Wild Sheep Creek. Bull Creek also contributed to this complex fan deposit.

Alluvial Fans

Alluvial fans are common features observed along the banks of the Snake River (Figure 34). Alluvial fans are named both for their detrital composition (alluvium) and fan-like shape. Most alluvial fans in the canyon formed by deposition of detritus from cataclysmic rock, debris, and earth flows that originated along and in the tributary creek drainage areas. During this process a rapidly moving (as fast as 40 mph) debris-filled stream loses velocity and momentum as it reaches the apex of the alluvial fan, because the gradient of the channel is dramatically reduced at that point. The pre-existing main channel crossing the alluvial fan may be too small to hold the debris and hence overflows. The detritus subsequently spreads across the fan as a sheet flow or, alternatively, carves a new channel in the fan.

Many of the tributary streams in Hells Canyon suffered "blowouts" during the winter of 1996-1997. These were particularly common in the southern part of Hells Canyon along the Oregon side where Homestead, Herman, Copper, McGraw, and Spring creeks have boulder-filled stream beds and new layers on alluvial fans. Their affect on the floor of Hells Canyon Dam reservoir has not yet been determined.

BONNEVILLE FLOOD

G. K. Gilbert, in an 1890 paper, recognized results from the catastrophic Bonneville Flood in the Snake River Plain region of southern Idaho. Interest in the Bonneville Flood grew in the 1960 to 1980 interval when several maps and papers were published on the subject. More recently, Jim O'Connor (see Bibliography) reviewed previous work and reported the results of studies that focused on the flood's dynamics. According to O'Connor, the Bonneville Flood roared across the Snake River Plain and through Hells Canyon approximately 14,500 years ago. Lake Bonneville in Utah, at one time covering an area of about 51,530 km^2 (equivalent to an area that measures about 75 miles by 250 miles), discharged 4,750 km^3 of water over the divide at Red Rock Pass, Idaho. The pass lies between the Lake Bonneville basin and the watershed of the Snake River. The level of ancient Lake Bonneville was lowered more than 100 m (about 327 feet) during the flood. Peak **discharge** was approximately one million m^3 per second at the outlet point near Red Rock Pass. At Lewiston, Idaho, about 1,100 km downstream, the maximum disharge had decreased to about 0.6 million m^3 per second.

O'Connor divided the Snake River Canyon below Hells Canyon Dam into three segments. The segments, mean depth in meters (and in feet), mean discharge in 10^6 m^3/sec (and 10^6 ft^3/sec), and mean velocity in m/sec (and miles per hour) are described as follows:

1) Bills Creek to Pittsburg Landing (river miles 234 to 214); depth of 170 m (557 feet); discharge of 0.57 x 10^6 m^3/sec (20 x 10^6 ft^3/sec) and velocity of 13.8 m/sec (30.8 mph).

2) Pittsburg Landing to Salmon River (river miles 214 to 188); depth of 187.6 m (615 ft); discharge of 0.57 x 10^6 m^3/sec; and velocity of 15.9 m/sec (35.6 mph).

3) China Garden to Lewiston (river miles 176 to 141); depth of 114.4 m (375 ft); discharge of 0.57 x 10^6 m^3/sec; and velocity of 14.5 m/sec (32.4 mph).

These figures differ significantly from the maximum pre-dam flood flows of an estimated 100 x 10^3 ft^3/sec in the upper parts of Hells Canyon (compared to 20 x 10^6 ft^3/sec) and a velocity of 7 or 8 mph (compared to a velocity of 30.8 mph in the first segment). It must have been a spectacular sight—if only humans had recorded the catastrophe. A similar but much shorter-lived flood would roar down the canyon if the three present-day dams (Brownlee, Oxbow, and Hells Canyon), all filled to capacity, were to break in rapid succession.

The Bonneville Flood caused many depositional (Figure 35) and erosional features that can be observed along the lower walls of Hells Canyon. Outstanding Bonneville Flood deposits are exposed at Pittsburg landing. Other specific features that resulted from the Bonneville Flood will be pointed out in the accompanying geologic guide. The rushing, high-powered flood-

ISLANDS AND RAPIDS

Figure 35. Bonneville Flood gravels smooth the north side (on the left) of Sluice Creek canyon. These gravels were deposited under a current that had slowed dramatically as it flowed up the tributary stream. Notice the incised alluvial fan on the Idaho side of the Snake River.

waters eroded canyon walls, in places undercutting cliffs and in other places leading to increased slope angles, both of which led to slope failures and landslides. The best example of slope failure that was probably caused by the Bonneville Flood lies along the Oregon side of the river between Marks and Waterspout creeks where large-scale slumping occurred.

MAZAMA ASH

White to light beige Mazama Ash crops out near the surface of many alluvial fans and in protected areas along tributary streams throughout the Hells Canyon region (Figure 36). Charles Bacon, in a 1983 journal paper, thoroughly discussed the eruptive history of Mt. Mazama. A cone-building stage occurred about 420,000 to 50,000 years ago. The volcano erupted catastrophically 6,845 +/- 50 years before present (BP) to form Crater Lake in the Cascade Mountains of southern Oregon. (The +/- indicates the degree of uncertainty in dating carbon that was collected from within the ash, by the radiocarbon method). For purposes of this book I chose 6,850 years BP for the approximate time of the climactic eruption, knowing that the eruption may have occurred at any time between about 6,800 and 6,900 years BP.) The climactic eruption formed a **caldera** nearly 4,000 feet deep and 6 miles in diameter. Wind-blown ash extended more than 1,200 miles northeast from Mount Mazama and covered more than 350,000 mi^2 of the continent.

Mazama Ash is found in deep sea sediments far from the coasts of Washington, Oregon, and northern California. The ash was initially transported by the Columbia River to the deep sea Astoria Fan (like an alluvial fan except that the fan formed in deep water) within a few weeks to months after the eruption. Subsequently, turbidity currents spread the ash over thousands of square miles. The ash is very important to marine geologists for relative dating of other late Quaternary deposits that occur in deep sea sediments off the coasts of Washington, Oregon, and northern California.

RAPIDS

Rapids in Hells Canyon are very common (Figure 37). Between the Hells Canyon Dam and the Oregon-Washington border about 70 rapids have been named. Many small unnamed rapids further disrupt the river's tranquillity. In general, rapids form by three methods: 1) a sudden steepening of the stream gradient, 2) the presence of a restricted channel, or 3) the unequal resistance

Figure 36. A thick layer of Mazama Ash overlies the Rush Creek landslide deposit. The ash is thicker here than in most places in the Hells Canyon region.

Figure 37. Rapids at Squaw Creek before Hells Canyon Dam was constructed. These rapids were a challenge to rafters and many were washed overboard.

of successive rocks traversed by a stream. Most rapids in Hells Canyon occur either at the mouths of tributary streams or in stretches of the canyon where landslides, debris flows, and rock falls constricted or blocked the river's flow. The river gradient was thereby changed by an influx of coarse debris that the river could not carry away.

Most large rapids in Hells Canyon are found at the mouths of tributary streams. Coarse debris is brought into the river during catastrophic runoff, either during a period of heavy rainfall, rapid snow melt, or after a landslide. In all cases, coarse detritus is brought down the stream in a rushing mud- and rock-laden torrent, sometimes referred to as a blowout (or **waterspout**). Because of an increase in its power (or capacity), the stream actually becomes a water-saturated debris flow that can carry large boulders and trees. When a boulder-laden tributary stream dumps boulders into the river, the river may lack the power to carry the larger boulders downstream. Consequently, the larger boulders stay in the river near the mouth of the tributary stream, diverting the channel and locally increasing the river's gradient. The river can gradually (for example, over a period of several decades) erode and break up the boulders, thereby reducing them to a size that can be transported by the river during high water.

SHINY ROCKS AND BOULDERS

A very curious feature along the banks of the Snake River throughout most of lower Hells Canyon is the proliferation of shiny rocks and boulders. The shiny rocks can generally be observed near water level, but in places they occur as many as 20 feet above the present water level. In general, there are two types of shiny rocks: one type is shiny because of a sand-blasting effect; the other type is shiny because secondary minerals were precipitated onto the rocks' surfaces.

The first type of shiny rock displays the minerals and textures beneath its shiny surfaces. An analogy can be drawn to small shiny rocks that are sold in rock shops; those shiny rocks are made by constantly turning (tumbling) them in a solution of water and fine-grained abrading material such as corundum. Along the banks of a river, this first type of shiny rock is formed by sand and silt abrasion during high water.

The second type of shiny rock is either black or dark brown and has a metallic luster. These shiny rocks are very similar to those displaying a *desert varnish* that is commonly observed on rocks and rock pavements in desert regions. Studies of desert varnish show that there are successive thin layers or coatings of detrital and precipitated minerals on the rocks. Most layers or coatings are composed of clay, fine quartz grains, and metal oxides such as iron (generally hematite) and manganese (generally birnessite). These metal oxides contribute to the black and brown colors. Close examination of the rocks under a microscope or high-powered magnifying glass show that interstices and fractures in the rocks are commonly filled with silt grains and often with sand. The metallic coatings are precipitated from solution. The fact that the shiny rocks and boulders in Hells Canyon are found almost exclusively near river level suggests that the metallic layers form during, as well as directly after, high water flows. The shiny black rocks farthest above the river probably formed before the dams were built. Those coatings that seem to be actively accreting near water level are associated with the present fluctuations of the river.

CHAPTER 3

Geologic Guide Between Hells Canyon Dam and the Mouth of the Grande Ronde River

INTRODUCTION

In this chapter, I discuss the geology between the Hells Canyon Dam and the mouth of the Grande Ronde River, an area that is accessible mostly by boats, horses, and hiking. In Chapter 4, I explain the geology from the Oxbow Dam to the Hells Canyon Dam along the Idaho Power Company road in Idaho, an area that can be seen from an automobile.

For purposes of explaining the geology in this chapter, I divided the canyon below Hells Canyon Dam by river miles into ten segments. The river miles that I use in both the text and on maps are consistent with two publications: (1) *U. S. Army Corps of Engineers (1988) Navigation Charts, Snake River, Lewiston to Johnson Bar* ; and (2) the U. S. Forest Service, Hells Canyon National Recreation Area brochure, *The Wild and Scenic Snake River.* The river miles given are the miles above the confluence of the Snake and Columbia rivers.

I begin this chapter at mile 247 near Hells Canyon Dam and end it just below the mouth of the Grande Ronde River at mile 168.5, a total distance of 78.5 miles. Each segment heading denotes the area covered and the miles above the confluence of the Snake and Columbia rivers, and in parentheses ( ), the miles below Hells Canyon Dam. I chose to traverse the canyon from the dam northward because of the ease in describing the geology. If the reader is going up the river by powerboat, rather than down the river, then begin at the mouth of the Grande Ronde River and interpret by river miles and geography. Alternatively, wait until the powerboat is returning downstream to study the geology.

Geologic maps are included in order to complement the written discussion. These geologic maps were compiled at an approximate scale of an inch to the mile, but some maps have been reduced to fit the page. Latitudes and longitudes, scales, and explanations are given on each geologic map. The contour interval on each map is 1,000 feet. A general explanation (Figure 38) shows all of the units that are included on the geologic maps. Parts of the general explanation are used on each map, depending on the units traversed within that particular segment.

The geologic maps in this book are the only recent maps published. A map at 1:250,000 scale was published by the author in 1974, but more work has since been done in the area and many interpretations have changed. The

Figure 38. Explanation of rock units and symbols for all geologic maps. Units that crop out in a particular segment are repeated in the explanation for each geologic map.

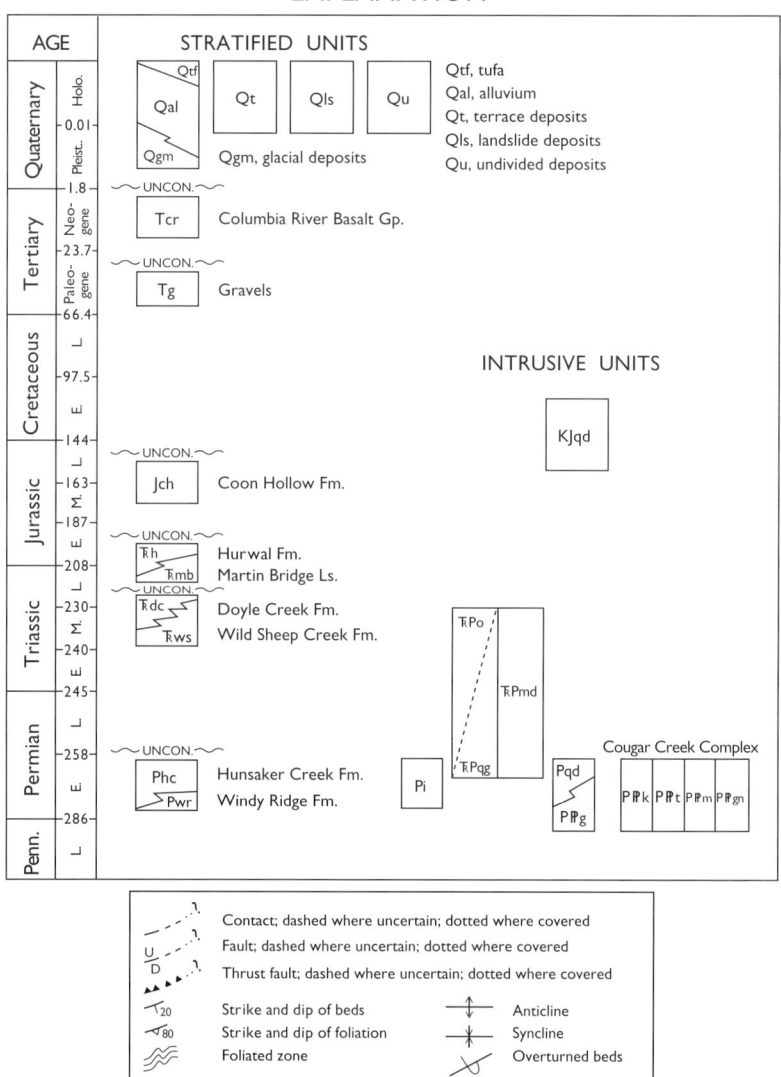

maps included in this book also will be changed as new information becomes available. I have never compiled a geologic map that I consider to be actually finished. I suspect that most geologists feel the same way.

Before your trip into Hells Canyon, I suggest that you color the maps using colored pencils. Match the color of each unit in the general explanation (Figure 38) with the units on the maps. For the stratified units, geologists generally use blues for Paleozoic rocks, greens for Mesozoic rocks, and browns and oranges for Tertiary rocks. Yellows denote Quaternary sediments. Intrusive units of all ages are generally colored in reds and pinks. I generally color the Columbia River Basalt Group a hue of lavender.

SEGMENT 1 [HELLS CANYON DAM TO BERNARD CREEK—MILES 247 (0) TO MILE 235 (12)]

Segment 1 (Figure 39a,b) traverses the most rugged part of Hells Canyon below Hells Canyon Dam. This segment shows Hells Canyon at its best:

rugged cliffs, cascading tributary streams, dynamic rapids, and a sense of the cataclysmic forces that raised (and are raising) the rocks along faults. The remarkably straight courses of the river in this segment follow faultlines; in fact, faults guide the river's course in many parts of the canyon. Some faults are ancient pre-Cenozoic structures, but others are younger than the Miocene flows of the Columbia River Basalt Group. I suspect that a few faults are currently active. An active fault is defined as having generated an earthquake in historic time or as having geologic evidence (such as offset sediments that are well dated) somewhere along its trend that the fault has moved at some time within the last 10,000 years.

Almost all of the rocks near river level in Segment 1 are part of the Wild Sheep Creek Formation. Hells Canyon Dam was built within westward-dipping strata of the Wild Sheep Creek Formation. The type area for the Wild Sheep Creek Formation is in the Wild Sheep Creek and Bull Creek drainage areas near mile 241.5. The Wild Sheep Creek Formation along this segment is composed primarily of volcanic flow rocks, coarse-grained volcanic breccia, and sandstone. Here and there, dikes of basalt and basaltic andesite, some of which may have been feeder conduits to the lava flows, cut the layered rocks. Limestone occurs but is very rare in this part of the canyon; north of the Snake River-Salmon River confluence, however, limestone is more common within the formation. It is easy to imagine that the river has cut through the sides of an ancient underwater volcano.

The launch site below Hells Canyon Dam (mile 247) is near the mouth of Hells Canyon Creek. Both the canyon walls and the stream channel of Hells Canyon Creek are very steep. A climb, either along the creek bed or to the top of the canyon wall, is not recommended for amateur hikers. Some hikers have been stranded and required rescue; other hikers have been injured. The visitor's center, presently managed by the U. S. Forest Service, was built along the south side of Hells Canyon Creek. The site may have been poorly chosen: a blowout down Hells Canyon Creek could destroy or engulf the structure.

A steep-walled crevasse, easily observed from the visitor's center on the north side of Hells Canyon Creek, marks the location of a deeply weathered and eroded dike of the Columbia River Basalt Group. It is probably a crystallized magma conduit or feeder to one of the overlying Miocene lava flows. Other straight, deeply weathered and partially eroded crevasses and lineaments somewhat similar to this can be observed in several places along Hells Canyon. Some of these lineaments can be identified by a change in either vegetation types or patterns. Vegetation changes are caused by differences in soil type and moisture retention. The difference in relief between the dikes and the surrounding rocks is caused by the relatively greater propensity for the dikes to weather and erode.

In Idaho, a hundred yards or so north of the boat launch site, a steep rock overhang marks a potential rockfall and avalanche (Figure 40). The flat and scoured surface near that menacing overhang, and the debris that has gathered below it, attest to an earlier rockfall.

The mouth of Deep Creek lies just below the dam in Idaho. Rocks in the creek bed are testimony to the many rock types transported by the stream. Deep Creek and its tributary streams drain most of the southern Seven Devils Mountains and traverse a wide variety of rock types, including Permian volcanic rocks of the Hunsaker Creek Formation, quartz diorite of the Cretaceous Deep Creek stock, some Permian and Triassic quartz diorite and gabbro plutons, and the Triassic Wild Sheep Creek Formation. About one mile

SUMMER, 1967. *I worked with Dr. Max Pavesic and his students from Idaho State University by identifying some of the rocks in their collection of Indian artifacts. During excavations on the alluvial fan of Hells Canyon Creek, they had unearthed a coffin that contained a skeleton with most of its bones still intact. Looters had stolen the skull. Weathered boots were attached to his feet and manila rope tied his ankles. To my knowledge, the remains were never identified. However, according to Carrey, Conley, and Barton (1979, p. 154) he had been dead for sixty to eighty years. I've often wondered if he were a revenuer? A rustler? A thief? A claim jumper? Or was his murder the culmination of an argument over a woman, a poker hand, or a jug of moonshine? Hells Canyon holds many other mysteries that may never be solved.*

Figure 39a, b. Geologic maps of Segment 1. Launch site for float boats and the U. S. Forest Service visitor's center are at about mile 247 (just below Hells Canyon Dam). North arrows, and a one-mile scale are given. The contour interval is 1,000 feet. See Figure 38 for explanations of all symbols used on the maps.

Geologic mapping completed by the author.

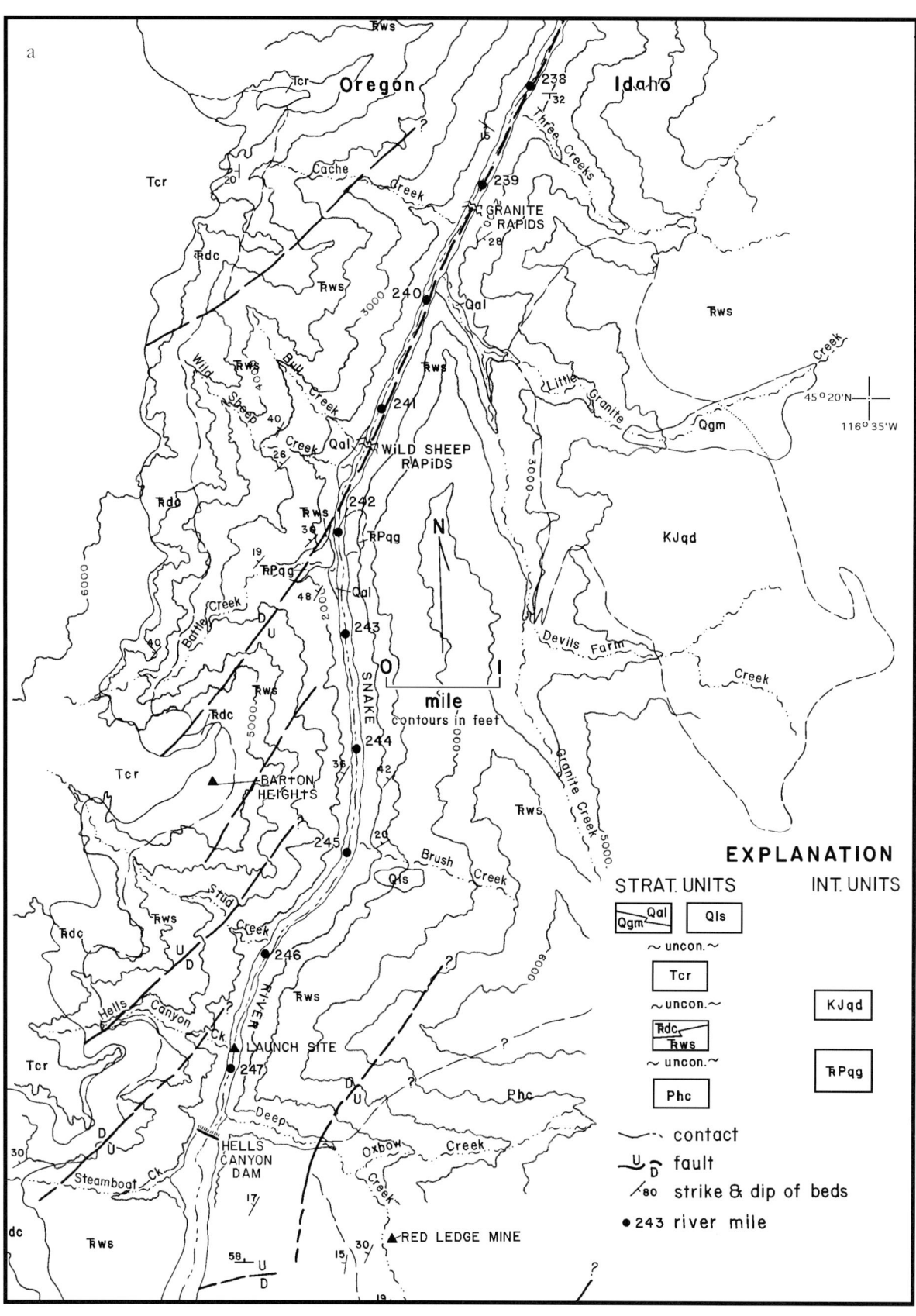

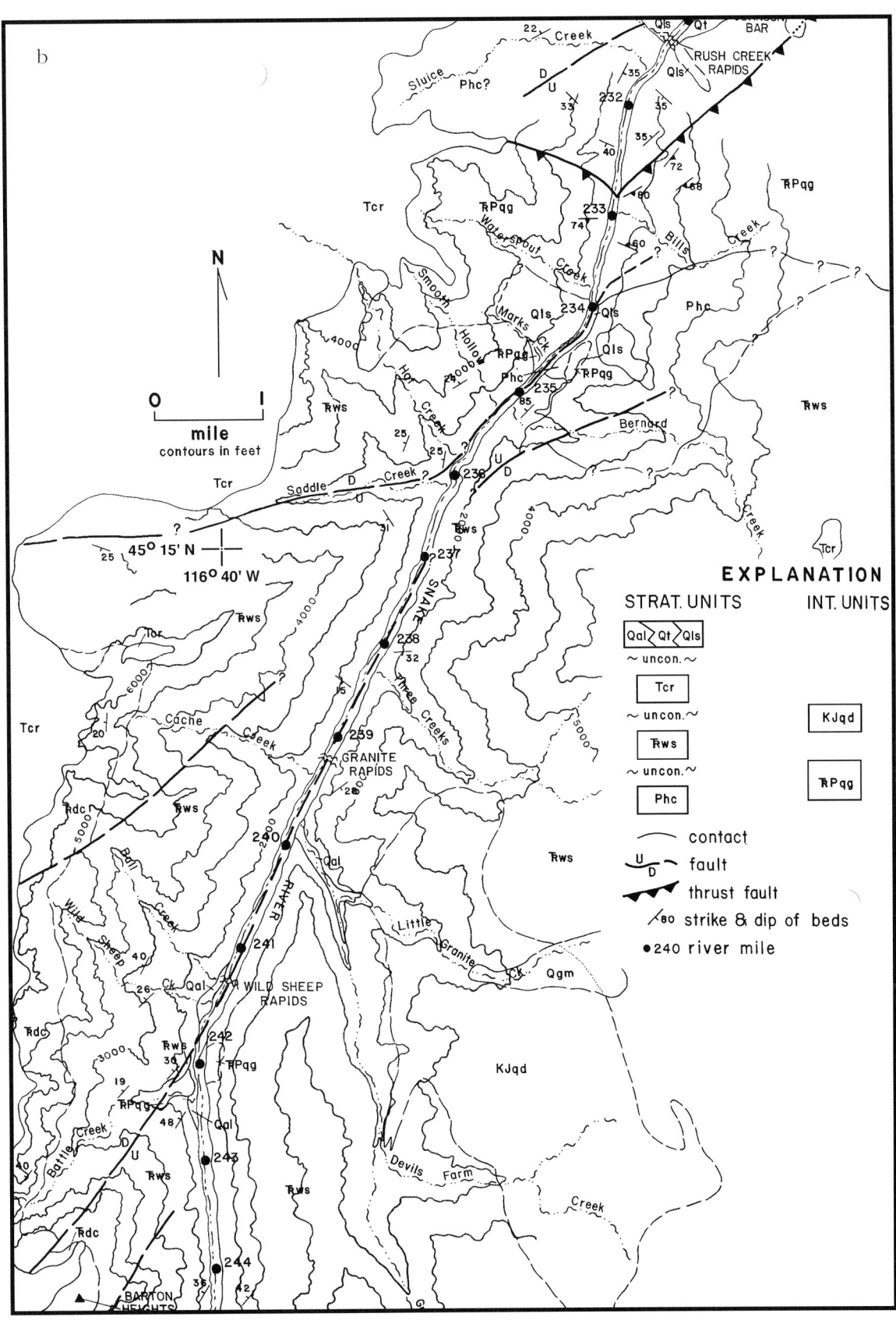

Figure 40. An imposing rock mass hangs menacingly over the Snake River near the boat launch site below Hells Canyon Dam.

above its confluence with the river, Deep Creek cuts through a fault that separates Permian from Triassic rocks. The unconformity between the dark green and rusty red rocks of the younger Wild Sheep Creek Formation and the light green, gray, and dark maroon rocks of the older Hunsaker Creek Formation occurs along the west wall of Deep Creek Canyon and can be observed about 500 yards west of the Red Ledge Mine. Fossil brachiopods were collected in rocks of the Permian Hunsaker Creek Formation about 300 feet below this unconformity.

The Red Ledge Mine (prospect) is approximately 2.5 miles above the confluence of Snake River and Deep Creek (Figure 39a). This mine has belonged to several different companies during the last 50 years. One major disadvantage to exploitation has been the difficulty of transporting the ore to a mill. The steep walls make road building an expensive challenge. And now the prospect is completely surrounded by a wilderness area, which means that ore from this mine may never be shipped. A portal to the main tunnel, several subsidiary portals, a yellow and red ore dump, and hundreds of feet of drill cores mark the site. The Red Ledge prospect is a zinc-copper-silver massive sulfide deposit. The mineralization is caused primarily by volcanic and tectonic activity that created an extensive stockwork system that was subsequently altered by hydrothermal (hot water) mineral-bearing solutions. Mine **tailings** are leaking some caustic metals and other substances into Deep Creek; the toxicity of these materials is unknown and hikers should fill canteens above the mine workings rather than below them. The mine is approximately 6 miles by trail from Eagle Bar, which is along the Idaho Power Company (IPCO) road about 2 miles south of the dam. Eagle Bar itself is now covered by the reservoir water, but a wide place along the IPCO road marks the location of buildings (now removed) that were constructed above that bar. A trail can be seen along the canyon wall leading to an old warehouse. I've hiked to the Red Ledge prospect several times from Eagle Bar. Although it is possible to hike there and back in one day, I don't recommend it. Instead, plan to stay overnight near the mine or somewhere along the trail.

Spectacular maroon, gray, and black rhyolite dikes can be observed north of the mine. These dikes were part of a Permian volcano. The rocks have high silica contents, relatively high amounts of sodium, and low potassium percentages. The high sodium and low potassium contents are opposite from unaltered rhyolite that is encountered in most environments. The rocks are referred to as quartz keratophyres and are probably the result of magmas, already low in potassium, coming in contact with sea water and thereby concentrating more sodium into the liquid rock.

This segment of Hells Canyon offers spectacular views (Figure 41) and visitors realize that they have seen the canyon at its best. At mile 245, the river elevation is about 1,420 feet above sea level and lies 4,300 feet below Barton Heights (1.5 miles west of the river), more than 5,000 feet below the western rim of the canyon, and 6,600 feet below Granite Mountain, which is 3.5 miles east of the river in the southwestern Seven Devils (Figure 42).

Near mile 246, look west into the rugged and steep canyon of Stud Creek where rocks are predominantly volcanic flows and volcanic breccias of the Wild Sheep Creek Formation. Waterfalls and other steep parts of the creek channel illustrate that the creek is not in equilibrium; downcutting by the stream has not kept up with uplift of the rocks. The creek is trying to reduce its gradient, but the drainage area is too small to even provide sufficient water for maintenance of the canyon. You might think of Stud Creek as a

Figure 41. Dry Diggins Ridge rises high above the Snake River near Bernard Creek in Idaho.

Figure 42. Snow-capped peaks of the Seven Devils Mountains form the eastern rim of Hells Canyon in Idaho.

ACCORDING TO *Carrey, Conley, and Barton (1979), Ralph Barton settled near Battle Creek in 1906 and subsequently found wire gold in the rocks along the south side of the creek. He staked a claim, but sold out to the Emnaha Gold Mining Company. The company built a sawmill, two bunkhouses, blacksmith shop, mill building, and tool shed. Workers packed in the materials by horse from the Imnaha River and then hauled the supplies over the top of the ridge. The ore ran out in 1916 and the mine was abandoned. Mr. Barton must have been a good prospector.*

Figure 43. Outcrops of the Wild Sheep Creek Formation in Bull Creek canyon. The Wild Sheep Creek Formation is the dominant rock unit in Hells Canyon (Frontispiece for Chapter 1, page 8).

starved stream; it is water starved and unable to cut the channel fast enough to keep up with the rising land. Many of the creeks in Hells Canyon are water starved, similar to Stud Creek.

The western rim of Hells Canyon peeks out here and there way above river level and, even from the river, one can see the dark horizontal layers that distinguish flows of the Columbia River Basalt Group. These flows can be seen more clearly downstream near Rush Creek where the western rim near Hat Point is visible from the canyon floor. Basalt flows of the Columbia River Basalt Group do not occur along the eastern rim because they have been eroded off the pre-Cenozoic rocks. Before late Cenozoic uplift, the horizontal flows of basalt extended continuously across what are now the Seven Devils Mountains and Hells Canyon; the lava flows were eroded off as uplift along faults continuously raised the rocks on the eastern side of the canyon higher than those on the western side. The same flows observed along the western rim (elevation around 5,000 feet) would be at an elevation of at least 10,000 feet in the southern Seven Devils Mountains if they had not been eroded. It is apparent that one or more faults have lifted the Seven Devils Mountains as much as a mile since the basalt flows poured over the ancient landscape.

Near mile 242, a small plutonic body of quartz diorite (most likely of Triassic age) crops out near the mouth of Battle Creek and along the lower part of the canyon in Idaho. I suspect that the gold found by Ralph Barton was near the contact between the quartz diorite and the older Triassic Wild Sheep Creek Formation.

When I mapped the Battle Creek area in 1967 some of the milling equipment still could be seen. Subsequently, I learned from Forest Service personnel that most of the equipment was buried during a 1978 blowout. I haven't been up that trail since 1967 so I can't comment on its present condition, but I do recall that the hike didn't require a very strenuous climb. Near the river, diorite and quartz diorite crop out near the trail. Farther up the trail rocks of the Wild Sheep Creek Formation are traversed. At about 4,800 feet elevation along that trail, reddish brown and maroon rocks of the Doyle Creek Formation are well exposed.

Bears are abundant in this part of Hells Canyon so keep your eyes open along the trail and, if camping, be sure to safely stow your food before turning in for the night.

An excellent example of an alluvial fan (Figure 34) occurs at the mouth of Wild Sheep Creek, just above the very rough rapids that have the same name. This fan is actually a combined Wild Sheep Creek and Bull Creek alluvial fan. The observation point for scouting Wild Sheep Creek Rapids is on a very thick and poorly sorted accumulation of angular-shaped boulders, cobbles, and finer-grained debris that make up the alluvial fan. These characteristics are typical of alluvial fan sediments in Hells Canyon.

The Bull Creek and Wild Sheep Creek drainages are the type area for the Wild Sheep Creek Formation (Figure 43). If camping on the alluvial fan, put on your boots and hike across Bull Creek and then up the north side of its canyon wall. Here, the Wild Sheep Creek Formation consists of volcanic breccia, **pillow lava**, sandstone, and limestone. Some of the breccia beds are more than a hundred feet thick. Each breccia bed is the result of one debris flow that had cascaded down the flank of an underwater volcano. In thinly bedded silty and carbon-rich limestone, flat clams of the genus *Daonella* can be collected from along some of the bedding planes. These fossil clams are characteristic of the Wild Sheep Creek Formation and are late Middle Triassic (Ladinian) in age (about 235-230 million years old).

The thick volcanic formations in Hells Canyon (Wild Sheep Creek Formation), Vancouver Island (Karmutsen Formation), and southeastern Alaska (Nicolai Formation) have the same fossil assemblages, indicating the same Middle Triassic age. Furthermore, Triassic rocks from each of these widely separated formations have the same paleomagnetic inclination (formed at about 20° N. Latitude) suggesting northward transport of several hundred miles since the rocks formed. Many geologists believe that these three areas (Hells Canyon, parts of Vancouver Island, and some of southeastern Alaska) were once geographically close together in the Middle Triassic and were part of an offshore microcontinent (or group of islands) called Wrangellia. According to this model, plate tectonic forces fragmented the landmass and subsequently transported the fragments to different areas along western North America. The chemical compositions of the basaltic lava flows, however, are somewhat different. Lava flows in Hells Canyon are related to island arc volcanism, whereas those in both the Karmutsen and Nicolai formations have chemical compositions that are closer to an oceanic plateau origin; this means that the lava flows in the latter two formations may have erupted above a deep mantle plume or **hot spot**. It is possible that the plume or hot spot initiated the subduction process, thereby leading to island arc volcanism that was responsible for lava flows in the Wild Sheep Creek Formation.

Lava flows and associated rocks in the Wild Sheep Creek Formation were erupted and eroded from volcanoes that formed as islands and seamounts in an island arc, similar to the volcanoes that now occur in the western and northern Pacific Ocean (for example, the Tonga, Mariana, and Aleutian island arcs). Most of the flow rocks are basaltic andesite and basalt. Chemical analyses show that the flow rocks are tholeiitic (for petrologists, this means that the flow rocks are iron-enriched, have nearly flat rare earth-element patterns and contain relatively low amounts of potassium). The rocks have suffered low- to medium-grade greenschist facies metamorphism; in some rock samples, the original pyroxene (augite) and calcium-rich **feldspar** (labradorite) crystals have been transformed to chlorite, epidote, and sodium-rich feldspar (albite). The degree of metamorphism, and therefore the resultant metamorphic mineral assemblages, vary somewhat throughout the formation, depending in part on the proximity to younger plutonic bodies and, in part, to the probable depth of burial that occurred during metamorphism.

Hells Canyon in this area is characterized by remarkably straight river segments, particularly between Brush Creek and Wild Sheep Creek (3 miles long) and between Wild Sheep Creek and Saddle Creek (nearly 6 miles long). These straight segments are most likely fault controlled, meaning that the river followed zones of weakness caused by the faulting. Factors other than faults may be involved in guiding the river course through this part of the canyon. For example, there are places in tributary canyons where streams were cut along dikes of the Columbia River Basalt Group that, when weathered, are relatively softer than the enclosing metamorphic rocks and consequently much easier to erode.

A trail that originates about a mile south of Brush Creek on the east side of the river in Idaho can be followed all the way to Pittsburg Landing, a distance of about 30 miles. A little farther downstream in Oregon, near Cache Creek, another trail begins that will take the hiker more than 26 miles to Pittsburg Landing. For courageous hikers in good physical shape, consider

HOWARD BROOKS and I mapped in the Battle Creek area in 1967. We camped on the lower bar and hiked up the Battle Creek trail the next morning. We pounded on rocks here and there, looked at the mining equipment, stuck our heads in the mine portal, and had a great time. The trail crossed a tributary stream about a mile from the river. We dropped our packs, got on our knees, and bent over the stream for a drink. A branch snapped about 10 yards upstream. Startled, we looked in that direction and saw a black bear gazing at us from behind a fallen tree. I immediately thought "Oh my gosh, maybe she has cubs," grabbed my hammer, and began running up the trail away from the bear. Howard was right behind me. I jumped on a rock ledge above the trail and threw rocks toward the bear, but it had disappeared. We never saw that bear again. It must have been just as surprised as we were because it ran in the opposite direction.

leaving the river at Granite Creek and hiking the approximately 16 miles to Windy Saddle in the northern Seven Devils Mountains of Idaho. The trail splits about a mile from the river; one trail leads up Little Granite Creek and the other parallels main Granite Creek and then cuts east up Devils Farm Creek. At the top of each of those trails, follow the loop trail either direction around the Seven Devils Mountains. I've hiked from the Snake River to Windy Saddle via the Little Granite Creek trail in two days with a heavy pack. A person in very good physical condition should be able to hike it in one long day. The elevation difference is about 6,000 feet.

Granite Rapids (near mile 239) is one of the most rugged and dangerous rapids in the canyon. At low water it is especially technical due to an abundance of boulders and the presence of a pool beneath a large boulder near the Idaho shore. I've seen rafts flipped in Granite Rapids and several jet boats have been lost in the large pool.

The west border of an Early Cretaceous quartz diorite pluton lies parallel to the river about a mile east of the mouth of Granite Creek in Idaho. White to light gray boulders scattered on the fan and small flood plain near the mouth of Granite Creek were eroded from that pluton. It is particularly well exposed in Little Granite Creek. The contact with the older rocks runs nearly north-south and is in part fault controlled. The pluton was dated radiometrically by the **K-Ar method** (two minerals, **hornblende** and biotite were used for the radiometric dates) and is about 115 million years old—the youngest quartz diorite pluton thus far mapped in Hells Canyon. This pluton is one of many Jurassic to Cretaceous plutons that are widespread throughout eastern Oregon. Best examples of these Jurassic to Cretaceous plutons are the Wallowa and Bald Mountain batholiths in eastern Oregon. In places, some of the Jurassic to Cretaceous plutonic bodies stitched the Blue Mountains Island Arc terranes together, beginning about 150 to 140 Ma. The plutons are neither tectonically deformed nor metamorphosed, thereby indicating that most of the large-scale deformation and metamorphism within the older rocks and terranes of the Blue Mountains had ended by approximately 150 Ma.

CARREY, CONLEY, *and Barton (1979) reported that Dave Hittsley was the first European to live along Granite Creek. Martin Hibbs followed Hittsley. The old home site is nearly a mile up the trail—an easy hike from the mouth of Granite Creek. The Hibbs family brought cattle there in 1902 and filed a homestead entry in 1911. Martin Hibbs died at his Granite Creek ranch on June 24, 1935, likely murdered by an addled prospector named Joe Anderson. Anderson, who Hibbs had asked to care for the place during a trip to Riggins, also died at the scene. Investigators believe that Anderson killed Hibbs and then turned the gun on himself. It is possible that someone else killed both men. Like so many other stories in Hells Canyon, this one also remains a mystery.*

THE SNAKE RIVER *in Hells Canyon can be unforgiving. She has snatched life from many unwary or foolish adventurers. Thus far I have escaped, partly because of experience, but mainly by luck. In my early days of study I made some near fatal mistakes, one of which occurred in early June, 1968, when Hells Canyon offered me a watery grave.*

Dave White, a student in my petrology class at Indiana State University, was my assistant that year. When I asked for someone to accompany me to Hells Canyon, he pluckily raised his hand. Neither of us understood the hard work and dangers the summer would entail.

I bought a ten-foot Japanese raft, a 3.5 h.p. Evinrude motor, and a wooden rowing and lashing frame. We took off for Lewiston where we hired Floyd Harvey to take us up the river; we planned to float the Snake in 7 days from Granite Creek to Lewiston.

Floyd grinned when he nosed into the Idaho bank of the river below Granite Rapids. He warned us about the high water and erratic rapids, but Dave and I were confident. We piled our gear on the bank and made camp. In the early evening a pack train approached along the trail; a woman on a brown horse was leading three burdened horses. She carried a .22 caliber pistol on her hip and the first horse had a treadle sewing machine tied to the packsaddle. After a short greeting, we learned that she had been up Granite Creek with "her man" and was on her way out to Riggins. When asked about the pistol, she said it was for rattlesnakes. Dave and I hoped we didn't resemble snakes.

Next morning we arose at sunup, ate a hurried breakfast of instant oatmeal, hot chocolate, and canned grapefruit sections, and inflated the raft. I don't know how much water the

Idaho Power Company was spilling through Hells Canyon Dam but a large Ponderosa pine log floated past. Based on later experience, I estimate that the river was flowing over 60,000 cfs (cubic feet per second); normal flow in late June generally averages about 20,000 cfs. We loaded the raft with 7 days of rations, camping equipment, cameras, and other belongings. Everything possible had been dutifully wrapped in black garbage bags that were then tied to the boat's frame. When loaded, less than 6 inches of freeboard remained between the water and the top of the raft. We buckled our life vests, and while Dave stayed ashore and held the bow line, I paddled out and started the motor. Dave waded out and jumped into the raft.

We didn't go far before a wave breached the gunwales and washed Dave into the water. He bobbed along for awhile, holding the extra aluminum oar that we had tied to the raft. Then he crawled back into the boat, disgusted because he had lost a yellow baseball cap. We were convinced that the river was too high, so we motored back to the bank to wait until the water receded. While Dave stripped gear from the boat, I took off my shoes and waded out to a rock outcrop that was still above water. I stood watching the river for 10 or 15 minutes, wondering if I really knew what I was doing. Suddenly, Dave yelled. I glanced over my shoulder as he pointed to a spot near my feet. A rattlesnake was coiled about 5 feet away; obviously, it didn't like the high water either, but luckily was not in the mood for a fight. I hooked the snake with a long pine branch and sent it swimming.

The water level had receded a few feet by daybreak, so I convinced Dave that time was wasting and that we should try again. The water still ran at least 50,000 cfs. At Three Creek Rapids (mile 238; nine miles below Hells Canyon Dam) we shipped several gallons of water. As we neared Saddle Creek, the roar of the rapids was ominous, and for the first time I felt incompetent. We motored to the Oregon bank and, with a rope, lined the boat through the white water. It took nearly an hour to maneuver the cumbersome raft through the rapids. Afterwards, we were tired and I knew we wouldn't want to line through any more rapids.

At my urging, Dave walked along the trail on the Oregon side of the river while I guided the raft through the next few rapids. I knew that Upper Bernard and Waterspout rapids were dangerous, and I didn't want to take a chance on losing a good assistant. Needless to say, Dave was not an expendable piece of geological equipment. I motored down the river with Dave running along the trail beside me. I shut off the motor above each stretch of white water and restarted it in the quieter water below.

A large wave flipped the raft in Upper Bernard rapids. Dave said later that the raft just flew up into the air and was upside down when it hit the water. I first surfaced under the raft, caught a lungful of air, and then dived and resurfaced alongside the raft. I grabbed the extra oar that had been attached to the gunwale, pulled myself to the raft, and climbed on top of the bottom of the now upside-down boat. The gear that had been tied ontop the raft now served as a water-logged keel. I hung on until the boat dived into Waterspout Rapids. At high water, Waterspout Rapids has a straight chute down the middle, but large curling waves roll upstream near the base of the rapids. One of those waves knocked me off the raft and into the water again. Not easily discouraged, I reboarded and intertwined my fingers with the line around the raft's perimeter. I drifted along as Dave ran to keep up, but I slowly outdistanced him. Bills Creek Rapids were mild and I finally floated into an eddy above Sluice Creek. When Dave caught up, I threw him the long bow line and he held it while I waded over to the bank. We tried to beach the boat, but the gear and cumbersome keel defeated us and we were unable to pull it to shore. I certainly didn't want to abandon the gear, so I leaped back onto the upside-down raft and rowed out of the eddy and back into the current. Sluice Creek Rapids and the upper part of Rush Creek Rapids treated me fine, but the large rock in the lower part of Rush Creek Rapids was right in the middle of a chute caused by the exceptionally high water. About 5 feet of rock loomed above the white water and I hit it square; the boat buckled, my head glanced off the rock, and I ended up in the water again. Dazed and cold, I pulled myself back onto the boat and floated past Johnson Bar and into a breath-taking ride through Sheep Creek Rapids.

Floyd Harvey's camp was about 200 yards below Willow Creek. When I reached the wide part of the river across, and a little upstream, from his camp, the boat veered left into an eddy and started floating upriver. I was shivering and couldn't move—unbeknownst to me, I had hypothermia. Dave came running along the trail and was able to pull me off the raft and help me reach shore. He towed the raft as close as possible to the bank and tied it to a large rock.

I had hung on from Upper Bernard Rapids to Harvey's camp near Willow Creek, a distance of about 7 miles. The wet journey took a little more than an hour and a half (even counting the time spent trying to beach the raft at Sluice Creek); I often wonder how Dave kept up with me in tennis shoes.

We pulled the gear from under the boat, and I can still recall the water floating in the lens of my expensive camera and noticing that most of the bluing was gone from the pistol. We dried the gear along the bank that afternoon: a myriad of plastic bags, tents, sleeping bags, and identical cans without labels. From that time forward, Dave and I had a surprise every time we opened a can.

Next morning Dick Rivers, in the IDAHO QUEEN mail boat, saw us on the bank. The boat nosed into shore, and amid the clicking shutters of tourists' cameras, we loaded our still soggy gear onto the IDAHO QUEEN. Dick ferried us across the river to Floyd's camp where we once again set the gear out to dry and waited for Floyd's arrival.

A few years later Dave told me that the summer of 1968 provided good training for Army Ranger school, which he attended before a tour in Viet Nam.

A stratigraphic section of the Wild Sheep Creek Formation was measured along the north canyon wall of Saddle Creek (mile 236). Abundant quartz-rich sandstone near the top of that section indicates that older plutons (probably Permian) had been unroofed and were being eroded during the deposition of Middle Triassic sand. Most other sandstones in the Wild Sheep Creek Formation were derived from basalt and basaltic andesite source rocks and are either completely devoid of quartz or contain only small amounts. In places such as the southern Wallowa Mountains near Fish Lake, however, plutonic rock clasts in conglomerate beds are common in both the Permian and Triassic rocks. Their presence in those conglomerates indicates the complexity of the island arc and suggests that it was similar to some of the modern island arcs such as the Aleutians where plutons of several ages (Eocene, Oligocene, and Miocene) have been studied.

Consider leaving the river and hiking the trail that begins on the Oregon side between Granite Creek Rapids and Saddle Creek. This trail leaves the river approximately a mile past Cache Creek and climbs to an elevation about 1,000 feet above the river. The trail parallels the river at that elevation for about 2 miles and then descends into Saddle Creek. It is an easy trail down Saddle Creek to the river. (Watch out for rattlesnakes and poison ivy.) The trail traverses the Wild Sheep Creek Formation the entire way. Rocks are mostly breccia, sandstone, and black carbonaceous limestone. *Daonella* fossils were collected from black limestone that crops out along the trail. At Saddle Creek, instead of turning right toward the river, a hiker may choose to walk west to Freezeout Saddle (about 8 miles from the river) and from there either west to the Imnaha River or north to Hat Point.

Between Dry Gulch and Bernard Creek, look east from the river where Dry Diggins Ridge rises precipitously. The top of Dry Diggins Ridge stands more than 6,000 feet above the river (Figure 41). A fire lookout tower at Dry Diggins has been abandoned. It is a relatively pleasant hike (stay overnight near the lookout and bring water) from the Seven Devils campground near Windy Saddle to the Dry Diggins lookout, but don't try to hike from the river because rock faces are steep and crumbly and there is no trail. A sense of history permeates Dry Diggins Ridge and one thinks not only about the isolation of the temporary summer occupants of the lookout, but also about Jack Hastings who mined there off and on for twenty years. From the north side of the ridge, Johnson Bar hugs the banks of the river nearly 6,000 feet below (Figure 4). Dry Diggins Ridge is underlain mostly by rocks of the Wild Sheep Creek Formation.

SEGMENT 2 [BERNARD CREEK TO SHEEP CREEK— MILES 235 (12) TO 229 (18)]

The segment of Hells Canyon between Bernard and Sheep creeks is geologically diverse (Figure 44). Rocks along the river near Bernard Creek belong to the Triassic Wild Sheep Creek Formation. Subsequently, starting at about mile 234, the river traverses gabbro, diorite and quartz diorite plutonic rock outcrops that are faulted over stratified Permian (Hunsaker Creek Formation) rocks. At Sheep Creek the river leaves the Permian rocks and once again slices through the Wild Sheep Creek Formation. It isn't just the rocks that are of interest here; the river in this segment has cut through three landslides and through a wide terrace sequence. The terrace sequence was formed in part by the planing off of landslides and alluvial fans during the high-water flow of the Bonneville Flood and probably in part by flood-water smoothing after the landslide dam broke at Rush Creek.

The old cabin near the mouth of Bernard Creek has been partially restored by employees of the U. S. Forest Service. A visit to that cabin and a walk across the alluvial fan below it will give you some feeling for the ranchers' isolation in Hells Canyon and the difficulties they had in surviving, both physically and economically. They were able to awaken each morning, however, to the muffled roar of Upper Bernard Rapids, the roaring, whispering, and gurgling of Bernard Creek, the melodious songs of canyon wrens, and the soft coos of turtle doves. Walk up the small knoll that lies south of the cabin where archeologists recovered some of the oldest evidence (greater than 7,000 years old) in Hells Canyon for Native American habitation. In addition, pictographs on the walls of a small rock shelter near the north edge of the high terrace, which lies just north of the alluvial fan, may illustrate the Native Americans' high regard for this area.

Rocks near the mouth of Bernard Creek are volcanic breccia of the Wild Sheep Creek Formation. Strata north of the creek's mouth are steeply dipping; some are overturned. Permian strata are in fault contact with the Triassic rocks about half a mile up Bernard Creek. The area is not well mapped and I'm not sure about the relationship, if any at all, of this fault to the long thrust fault that I mapped farther north near Johnson Bar.

A spectacular slump and landslide changed the course of the river along its Oregon side between Marks and Waterspout creeks, beginning at about river mile 234.5 (Figure 45). The landslide is really a multiple slump, part of which gave way and cascaded into the river. Based on the presence of boulders high on the Idaho side, I suspect that the landslide temporarily dammed the river to a depth of about 300 feet. I further suspect that oversteepening of the canyon wall (by the Bonneville Flood?), possibly combined with shaking by one or more earthquakes, caused the slump. Topographic features in the bowl-shaped depression across from the slump (in Idaho) suggest that the river flowed around the slump-caused dam for some time before cutting its way through to make the present channel. River boulders in that bowl-shaped depression, high above the present river level, strengthen that interpretation. The slump debris crops out along steep slopes between Marks and Waterspout creeks. In places, percolating ground water oxidized some debris,

ANYONE WHO HAS *camped in Hells Canyon for any length of time has a bear tale. And the bears are more numerous now than they were in the 60s and 70s. So are cougars, lynxes, and bobcats. In 1990, my assistants and I had a memorable encounter with a bear when Kent, Julie, Linda, Paul, and I camped at Upper Dry Gulch campground (mile 237.5). Roy Lombardo of the U. S. Forest Service dropped us off the jet boat the day before and we planned to work out of the Dry Gulch campground for 2 or 3 days. We marched out in the early morning, a string of 5 robust adventurers. Kent split at Three Creeks to work along the trail, and the rest of us turned east and climbed the ridge south of the creek. A forest fire had jumped through this area, probably in the summer of 1986, and left many charred trees in our way. It was more or less a routine climb, and we reached an elevation about 2,000 feet above the river by 10 o'clock. We sat in a shady nook, drank water from our canteens, and nibbled on trail mix. As we arose and girded our backpacks, we heard a branch snap. Linda asked, "What was that?" "Just a bear," I joked. However, when I glanced up the hill no more than 30 feet away, a large cinnamon-colored bear poked its head around a rock. I yelled, "Bear. Get down the hill." Then I picked up my hammer and chased the bear. Lucky for us—me—the bear ran and a few minutes later we saw it crossing a meadow high above us, long golden hair rippling in the sunlight.*

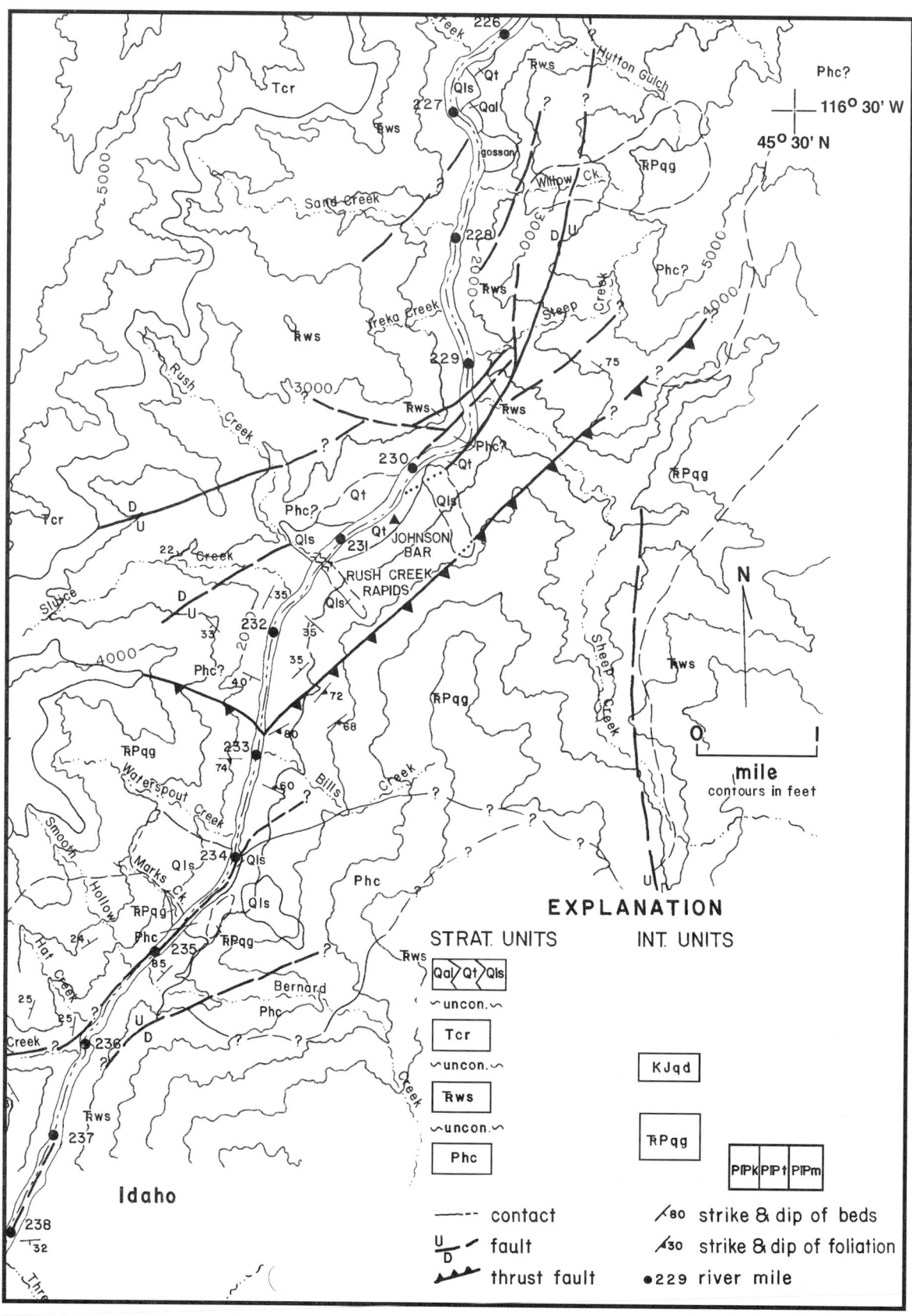

making brown-stained outcrops. In time, the slump will probably move and dam the river again (hopefully, not during your lifetime).

Evidence of erosion during high flood waters is carved in rocks along the foot trail on the Idaho side of the river, upstream from Waterspout Rapids. For a view of these features, beach your boat above the Waterspout Rapids and walk along the Idaho trail for about one half of a mile to see results of water scouring. For example, at about mile 234, **flute** marks and a deep rock-scoured hole show the tremendous cutting power of the river in flood. A large rock swirling around and around in a bedrock depression can actually erode holes. One of these holes, along the east side of the trail, is at least 6 feet deep with a diameter of about 4 feet.

The river traverses plutons that are Permian and Triassic (probably mostly Permian) in age between mile 234 and mile 233. The plutonic rocks are well exposed near the mouth of Waterspout Creek. Good outcrops also occur along the trail north of the creek and near Bills Creek on the Idaho side of the river. The rock types are diorite and quartz diorite; elsewhere within this complex body of plutonic rocks, gabbro and trondhjemite also are present. Along the river trail in Idaho, above and north of Waterspout Rapids, a border zone of the plutons contains sausage-shaped (**boudins**) quartz diorite **xenoliths** that are aligned parallel to the bounding thrust fault. One can comprehend, by closely studying these spectacular outcrops, the high temperature and pressure regimes that must have existed deep within the earth's crust when the rocks formed.

The thrust fault that bounds the plutons on the north side is not well exposed near the mouth of Bills Creek. From there, the fault trends northeast and I've traced it for more than 4 miles to a point north of Sheep Creek. From that point on, it has not been mapped. I presume that it is an old structure, probably of Late Triassic to Late Jurassic age. The fault places gabbro, diorite, and quartz diorite over the Permian Hunsaker Creek Formation. It is marked in places by a mylonite zone more than 300 feet thick. The best place to observe the mylonite zone is reached by climbing the eastern canyon wall to an elevation approximately 500 feet above Rush Creek Rapids (mile 231). Rocks of the Hunsaker Creek Formation were shattered and then recrystal-

Figure 44. Geologic map of Segment 2 in Hells Canyon. In this segment, the river cuts through the Triassic Wild Sheep Creek Formation, Permian Hunsaker Creek Formation, and Permian and Triassic (mostly Permian) gabbro, diorite, and quartz diorite plutons that were faulted over the Hunsaker Creek Formation.
Geologic mapping completed by the author.

Figure 45. Slump between Marks and Waterspout creeks, Oregon. Note the scarp that forms a steep headwall. Arrows show the movement.

CARREY, CONLEY, *and Barton (1979, p. 205-208) recount an interesting story about Si Bullock, a bachelor who built the rock house near the mouth of Bills Creek. Si came to Hells Canyon in 1912, built the 12' x 20' house, and lived off the land for about fifteen years. His story confirms the sustainability of Hells Canyon and its ability to provide for the inhabitants. I wonder if Si had a copy of Thoreau's Walden?*

lized to mylonite; mineral folia of quartz, chlorite, epidote, and feldspar parallel the fault plane.

I recommend a one-day hike to Hat Point and back from the mouth of Sluice Creek if you are in good physical condition and would like some challenging exercise. (Hat Point lies just off the left side of the map in Figure 44, west of the headwaters of Sluice Creek.) Dave White and I hiked to Hat Point in 1968 and were back in camp by 5:00 p.m. We left about 6:00 a.m. and ate lunch at the lookout. The trail up Sluice Creek to Hat Point is a steady climb. The rocks along the trail are mostly of Permian age all the way to the topographic bench where the first basalt flows are encountered, a little more than 3,000 feet above the river. From the bench to the top, the trail traverses flow after flow of the Columbia River Basalt Group. The hike is about 14 miles round-trip and the elevation difference is about a mile. Hat Point is nearly 7,000 feet above sea level whereas the river near its confluence with Sluice Creek is at about 1,300 feet. The hike is similar to climbing the south rim of the Grand Canyon from Phantom Ranch and returning the same day. Views along the trail to Hat Point are spectacular. In late spring and early summer, a herd of elk may be grazing along the high bench. Wild flowers alone make it worth the climb in late June or early July.

Gravels deposited by the Bonneville Flood cling to the north side of Sluice Creek about one half of a mile above the confluence (Figure 35). Similar gravels become more common in tributary canyons farther downstream. They are the result of sediment deposition from Bonneville floodwaters that were backed up (backeddies) into the side canyons; due to the decrease in water velocity and power, some of the sediment load was dropped. Rattlesnake Creek, which joins Sluice Creek about a mile from the river, is appropriately named. Based on my experience, I believe there are significantly more rattlesnakes in the Sluice Creek and Rush Creek areas than in any other part of Hells Canyon.

Rush Creek enters the river about one half of a mile below Sluice Creek. The geology at the mouth of Rush Creek foretells future events in the river canyon and its tributaries. The flat terrace, discernible about 400 feet above the river, is the top surface of a landslide deposit (Figure 46). The landslide (rock avalanche) originated a long distance above the river on the Idaho side, cascaded down the chute directly across from Rush Creek (Figure 32), and dammed the Snake River to a depth of nearly 400 feet. If the dam were unable to break before being overtopped, water may have been backed up more than 30 miles, almost all the way to Oxbow. After the temporary dam broke, Rush Creek cut a steep V-shaped canyon through the remaining debris on the Oregon side of the river. A house-size boulder observable above and south of the high terrace is composed of gabbro and was ripped from outcrops as high as 2,000 feet above the river in Idaho. The large boulders in and near Rush Creek Rapids were brought to the river in that landslide. When the dam broke, flood-waters smoothed some of the landslide debris and also the alluvial fans downstream and created wide terraces along the river. In one place north of Pony Creek, landslide floodwaters apparently cut into the higher Bonneville Flood terrace.

The exact age of the landslide is unknown, but in the canyon of an unnamed creek about 400 yards north of Rush Creek, stratigraphy tells the story (Figure 47). In that outcrop, landslide debris overlies large boulders that were rounded and deposited by the Bonneville Flood; alluvial fan debris and Mazama Ash clearly overlie the landslide debris. Therefore, the landslide

cascaded into the canyon between about 14,500 years ago (approximate age of the Bonneville flood) and 6,850 years ago (approximate age for the eruption of Mt. Mazama). There was sufficient time between the landslide deposition and Mazama Ash accumulation for a relatively thick sequence of alluvial fan debris to be deposited. Possibly, then, the landslide occurred closer to the time of the Bonneville Flood. The oversteepened and deeply weathered rocks high above the river in Idaho, between Bills Creek and Sheep Creek in Idaho, pose a future threat for more landslides. Just mix those unstable slopes with some wet years (and possibly an earthquake with a Richter magnitude of 4.0 to 5.0) to form another giant landslide.

Near Hat Point the Snake River canyon is deeper than the Grand Canyon of the Colorado *as measured along an east-west line from both sides of the river*. He Devil Mountain, the highest peak in the Seven Devils Mountains (Figures 48 and 49), is nearly 9,400 feet in elevation and lies just 6 miles east of mile 241 (near the mouth of Wild Sheep Creek). As measured from the top of the He Devil Mountain to the bottom of Hells Canyon, the elevation difference is approximately 8,030 feet.

Figure 46. Rush Creek landslide. This landslide deposit is nearly 400 feet high and formed when a slope of deeply weathered gabbro failed along the Idaho canyon wall and cascaded into the river.

Figure 47. Stratigraphic succession in the Rush Creek area consisting of, from bottom to top, Bonneville Flood boulders, Rush Creek landslide debris, local alluvial fan sediment, and Mazama Ash.

Figure 48. The Seven Devils Mountains viewed from a point near Hat Point, Oregon *(Color section, page 92).*

Figure 49. The Seven Devils Mountains near Sheep Lake, viewed from the north *(Color section, page 93).*

Figure 50. Landslide deposit at Johnson Bar. Notice the source area, now a basin, for the landslide deposits.

Figure 51. Columnar jointing in flows of the Columbia River Basalt Group in Idaho, across the Snake River from Asotin, Washington *(Color section, page 93).*

Johnson Bar, between miles 231 and 230, consists of a wide terrace and remnants of a landslide near the north end (Figure 50). The landslide has Mazama Ash on top so it is more than about 6,850 years old. And it is apparently younger than the Bonneville Flood because those flood-waters would have planed off the debris. Above the terrace, you may be able to see small scarps that mark the site of a potential landslide. In 1970, I drove 4 long metal stakes into rock and soil between the scarps; by the Summer of 1971, after a wet Spring, 2 of the stakes had been bent by downslope creep. There has been no appreciable movement since then, but most of that hillside is unstable and probably will move if there are successive wet years and gravity is assisted by a strong earthquake.

The beach on the north side of Johnson Bar has lost thousands of tons of sand since the upstream dams were built. In fact, almost all of the beaches along the river from the Hells Canyon Dam to the mouth of the Salmon River are sediment starved and sand is lost to the river channel during peak water discharge from the dams. Some of this sand has accumulated on the floor of the reservoir behind Lower Granite Dam.

About half way between the north end of Johnson Bar and Sheep Creek the river trail in Idaho passes through Permian rhyolite dikes that display spectacular **columnar** jointing. This is the only place along the entire length of Hells Canyon where Permian rhyolite dikes are easily observed. Other rhyolite dikes occur near the Irondyke and Red Ledge mines in the southern part of Hells Canyon within a mile or two of the river.

In the region, columnar jointing is generally associated with lava flows and dikes of the Columbia River Basalt Group (Figure 51), but near Johnson Bar the columns formed in rhyolite dikes. Columnar jointing has a propensity to form perpendicular to cooling surfaces. Where the ground is level, the columns are nearly vertical. Many columns, however, are horizontal and others are arranged at peculiar angles. If the ground is irregular, as in an arroyo or small canyon, the columns may form a curved or somewhat radiating configuration. Lava flows that fill a pre-existing lava tube (cave) may crystallize into columns that resemble the spokes of a wagon wheel.

A trail follows Sheep Creek for several miles on its way to Windy Saddle in the Seven Devils Mountains. This approximately 12 mile trail originates at an elevation of about 1,300 feet at the river and tops out around 7,600 feet at Windy Saddle. It is a challenging hike, especially on a hot summer day. Good outcrops of a Permian trondhjemite, mentioned in the preceding section, are visible along the trail about a mile from the river. The large boulders on the Sheep Creek alluvial fan, and in the stream itself, were eroded from that pluton. The trondhjemite was dated radiometrically using two different methods. The first was completed by Ted McKee (U. S. Geological Survey) who used the conventional K/Ar whole rock method in the early 1970s. Although the procedure and age determination were correct, the age of about 230 million years did not match the other field data and my geologic mapping. Therefore, more samples were collected in 1979 and zircon crystals were separated by Nicholas Walker (then at the University of California in Santa Barbara) who used the U/Pb method to determine that the trondhjemite crystallized about 263 million years. The discrepancy between the two methods is understandable because it is apparent from thin section studies that the trondhjemite pluton was metamorphosed after it had crystallized from a magma. I conclude that the 263 Ma year date signifies crystallization—about the same age as the Permian Hunsaker Creek Formation—and the 230 million years K/Ar radiometric date indicates the age of metamorphism.

A spectacular breccia separates Permian from Triassic rocks in outcrops that occur just north of the Sheep Creek trail about one half of a mile east of the cabin. This breccia marks the Permian-Triassic unconformity. Fragments of the Permian trondhjemite, mentioned above, occur in the breccia, thereby indicating that the trondhjemite was being eroded at the time of breccia deposition.

SEGMENT 3 [SHEEP CREEK TO HOMINY CREEK— MILES 229 (18) TO 223 (24)]

Segment 3, from 18 to 24 miles below Hells Canyon Dam, cuts mostly through the Wild Sheep Creek Formation (Figures 44 and 52). The rocks, however, are intensely sheared in the northern part of this segment and other rock assemblages could be involved, particularly the Permian Hunsaker Creek Formation. Not only have the rocks been crushed and shattered along multiple fault zones, but also many were mineralized by hot, mineral-charged fluids. These mineralized zones stand out as yellow- and brown-stained outcrops along the canyon walls.

Steep Creek is the first major tributary of the Snake River north of Sheep Creek. It is well named. The outcrops along the trail north of Sheep Creek in Idaho are volcanic flows, breccias, and sandstones of the Wild Sheep Creek Formation. In places, dikes with **porphyritic** textures cut the formation. The formation is coherent and not greatly deformed until the trail reaches Willow Creek; from there to Hominy Creek Rapids the rocks are much more deformed.

Sand deposits occur in many places along Hells Canyon. They commonly occur as high as 200 feet above the water. When sand was more abundant on the beaches, high-velocity winds often blew the sand uphill; on rare occasions, the sand deposits formed small dunes. A good example of these sand deposits can be seen along the Oregon side of the river near Yreka Creek.

THE ALLUVIAL FAN *at the mouth of Sheep Creek, according to Carrey, Conley, and Barton (1979, p. 216-222), was first occupied by William McLeod in 1884. He lived there until ill health drove him from the canyon to a town. Fred and Billy McGaffee bought the place from the County. Fred McGaffee and his family built the cabin that now sits on the alluvial fan near Sheep Creek. The U. S. Forest Service manages the property.*

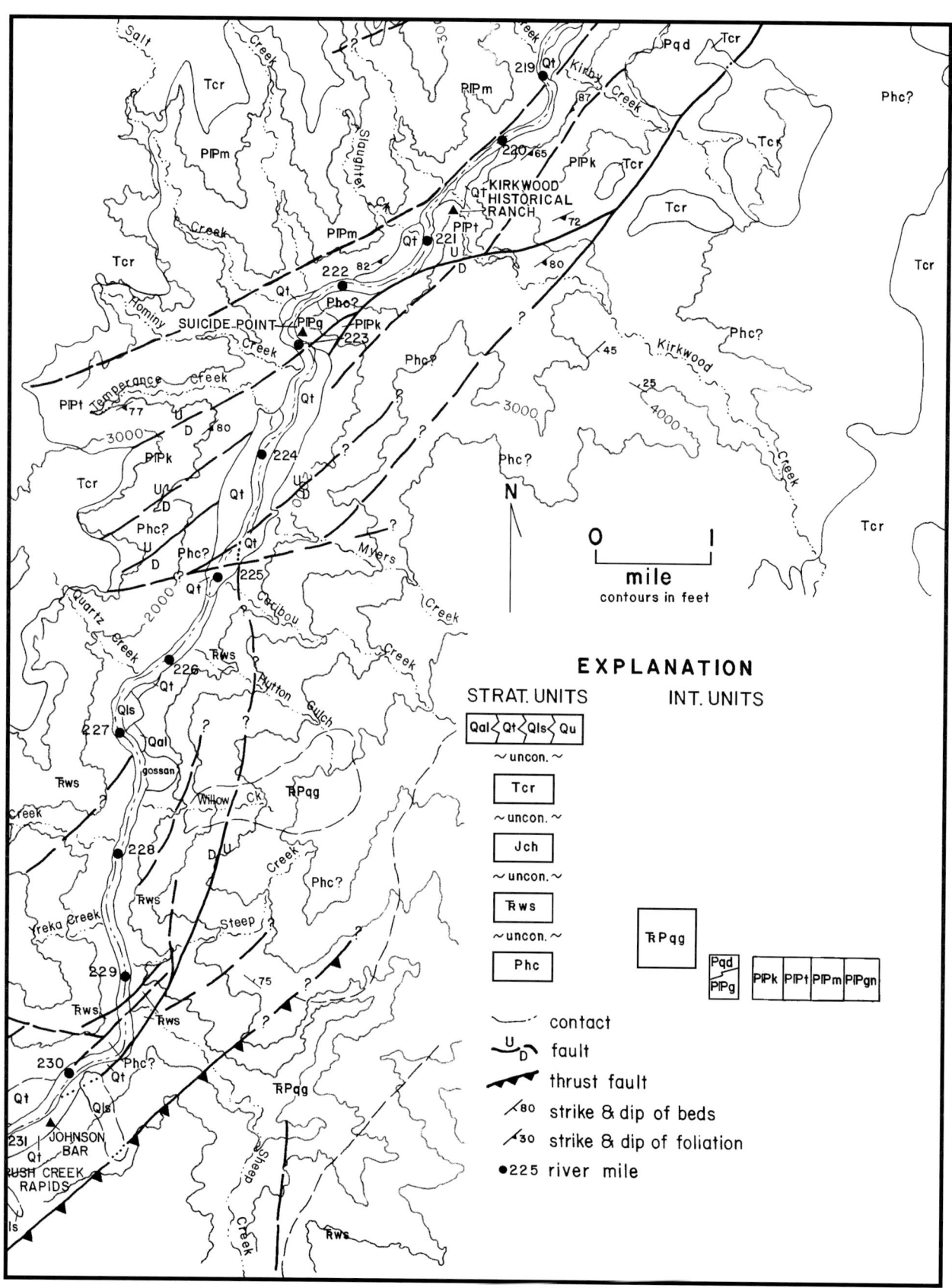

An imposing and unstable rock mass perches menacingly above the trail in Oregon at Eagle's Nest (Figure 53). This rock mass, like the one that towers above the river near the launch site below Hells Canyon Dam, may slide into the canyon during our lifetime. Don't pitch your tent under Eagle's Nest, and be sure to walk warily along the trail below it.

At Pine Bar, about 100 yards north of Willow Creek, the river opens into a spectacular vista. A small campground is located along the south end. A wide beach decorated Pine Bar until about 1971. The sand nearly reached the large rock that now juts out near the middle of the river. One could wade out on soft, clean sand to within 30 feet of the rock and then leisurely swim over to it. That beach is gone now, eroded away like most of the beaches along Hells Canyon above the mouth of Salmon River.

The ocher-colored and pine-studded outcrop above Pine Bar in Idaho (Figure 54) is a **gossan**. Gossans are mineralized zones, and as such they are a great benefit to prospectors because gossans are easily seen. Most gossans mark sulfide deposits that may contain metals such as gold, silver, and copper. The most common sulfide is pyrite (fool's gold), which consists of iron and sulfur. The iron and sulfur alter (oxidize) to ocher-colored and rusty-red rocks that contain secondary minerals such as hematite, **goethite**, quartz, and sulfates. This deposit also has been referred to as an **alum** bed. Alum is a common name for a wide variety of sulfates that form during the oxidation of sulfides. Close examination of the deposit indicates that parts of it occupy a crushed zone, related most likely to movement along a fault. Hot fluids (hydrothermal), probably derived from a crystallizing magma body far below, were transported along fractures to this level where abrupt changes in pressure, temperature, and water content allowed the sulfides to precipitate. The sulfide minerals were subsequently oxidized, thereby forming the ocher and rusty-red outcrops.

IN THE EARLY *evenings of the 1968 summer field season, Dave White and I often heard the muffled roar of Floyd Harvey's jet boat as he approached our camp. He'd drift up to within shouting distance and ask, "Hey, you guys hungry?" Our mouths watered, because that meant a steak dinner, a chance to sleep on a mattress, and beer hotcakes for breakfast. We always answered with an enthusiastic "Yes" and off we'd ride to his camp where several guests were waiting. After dinner Floyd started a fire on the sand bar and told stories about people who had lived and worked in the canyon, starting with the Native Americans and ending with the wheelbarrow woman. I'd tell them about the rocks and how I thought Hells Canyon formed. Floyd always ended our talks with a fervent plea for help in saving the canyon from more dams. The next morning Floyd would feed us hotcakes and take us back to our camp. Sometimes, we'd stick around and go fishing. I revere the canyon more because of Floyd. He contributed greatly to the preservation of Hells Canyon: a landscape free from dams, a wild and scenic river, a National treasure. An arsonist burned Floyd's camp to the ground on January 31, 1974. Floyd lost his livelihood and, more importantly, the river lost a devoted friend.*

Figures 52. Geologic map of Segment 3. See Figure 38 for the explanation of all rock units in Hells Canyon.

Geologic mapping completed by the author.

Figure 53. Unstable rock mass hangs precipitously over the trail at Eagle's Nest just north of Sand Creek in Oregon.

Figure 54. Gossan near the Willow Creek campground. Some river guides will refer to this as an alum deposit, but it is much more than that.

Figure 55. High Bar, as seen from the south near Pine Bar, was first formed by a landslide, and subsequently smoothed by Bonneville Flood waters *(Color section, page 94)*.

Figure 56. River terraces topped by alluvial fans near Temperance Creek fan.

A rock slide changed the course of the river just north of Pine Bar. This ancient rock slide forms the south end of High Bar (Figure 55). The high flat terrace on the north side of High Bar probably resulted from the deposition of Bonneville Flood gravels behind the landslide and the subsequent smoothing by waters of that flood. Hence, the slide occurred before the Bonneville Flood. The flat top (terrace), more easily observed from the north side of High Bar, is long enough to land small airplanes.

Another gossan occurs in the upper regions of Quartz Creek canyon (near mile 226) where red iron-stained outcrops caress the canyon walls. This gossan and the one at Pine Bar probably formed at the same time. A debris flow (blowout) cascaded down Quartz Creek in 1994, nearly taking out the picnic table. Blowouts like this should be expected at any time, particularly after an extraordinarily wet Spring.

Wide terraces line the river (Figure 56) between miles 225 and 223. These terraces originally were alluvial fans and landslide debris that had been brought down creek channels and sloughed off steep walls of the river canyon. Long after the fans formed, water from the Bonneville Flood, and probably other floods like the Rush Creek flood, smoothed off the older deposits; in addition, water was backed up behind Suicide Point, which provided a stricture in the canyon similar to a dam. This stricture caused the water velocity to decrease, thereby allowing suspended sediment to drop to the bottom of the temporary lake or reservoir. This added a small amount of sediment to the already planed-off terraces. Younger alluvial fans subsequently formed on the terraces, thereby creating a slightly concave upward surface. The high terrace in Idaho is called Big Bar. This "Big Bar" should not be confused with the "Big Bar" farther south near Allison and Eckels creeks that is drowned by the reservoir waters impounded behind Hells Canyon Dam.

Broken and shattered rocks along both sides of the canyon, near mile 224, indicate the chaos and convulsions that occurred deep beneath the surface millions of years ago. Not one fault, but rather a wide zone of faults (fault zone), runs through this part of the canyon. Small mine portals high on the sides of the canyon in Idaho testify to man's hunger for gold and his stubborn resolve to find it. Those holes were made by pick, shovel, and blasting powder. The inveterate and determined prospectors believed that either the next bite of the pick, the next shovel full of rocks, or the next explosion would yield the elusive "pot of gold."

SEGMENT 4 [HOMINY CREEK TO UPPER PITTSBURG LANDING—MILES 223 (24) TO 216 (31)]

Segment 4 cuts through about 7 miles of the Cougar Creek Complex (Figure 57). The Cougar Creek Complex is in part an igneous or plutonic "basement" to both the Permian and the Triassic stratified rocks in the Wallowa terrane. The crystalline rock (igneous and metamorphic) unit is thrust-faulted over Jurassic sedimentary rocks at Pittsburg Landing near mile 216. The assemblage represents rocks that formed near the middle of island arc crust. The older part (**amphibolite**, norite, and gabbro) is on the north near Upper Pittsburg Landing and the younger part (rhyolite, quartz diorite, diabase, and trondhjemite plus rare meta-sediments and lavas) is on the south near Temperance Creek. The mineral assemblages within rocks on the north part suggest that metamorphism occurred as deep as 10 km below the surface and that temperatures were greater than 500°C.

Almost all rocks in the Cougar Creek Complex are igneous and metamorphosed igneous rocks; original igneous textures remain in many of the rocks, but others have been greatly deformed and recrystallized during higher temperature and pressure conditions. Metamorphic rock types are gneissic amphibolite, amphibolite, hornblende and chlorite schist, **quartzofeldspathic** gneiss, and feldspathic gneiss. Mylonites are common along shear zones. Almost all rocks have a pervasive northeast-trending **foliation**. In places, the foliation is folded. Many foliation planes have nearly horizontal lineations, formed by elongate quartz and **amphibole** crystals, that suggest horizontal (strike-slip) movement. Boudinaged gneissic layers are common. Protoliths of the metamorphosed igneous rocks are norite, gabbro, diabase, diorite, andesite, quartz diorite, trondhjemite, and rhyolite, most of which crystallized as dikes. Rare undeformed gabbro and quartz diorite plutons cut the dike sequences.

ACCORDING TO *Carrey, Conley, and Barton (1979), Thomas Myers (Myers Creek on the Idaho side was named for him) lived a short distance up that creek and just above the trail. He shot and killed George Brownlee and wounded Wallace Jarrett on May 18, 1904 after they had argued over cattle and grazing rights. Apparently, Jarrett's and Brownlee's cattle had eaten lunch in Myer's garden, and that was the last straw. After Myers turned himself in and spent a day or so in the White Bird jail, Deputy Sheriff Seay tried to get him out of town because of a lynch mob. As Seay hustled Myers out of town, about 30 men approached them. The men took Myers away from Seay and hanged him from a tree limb.*

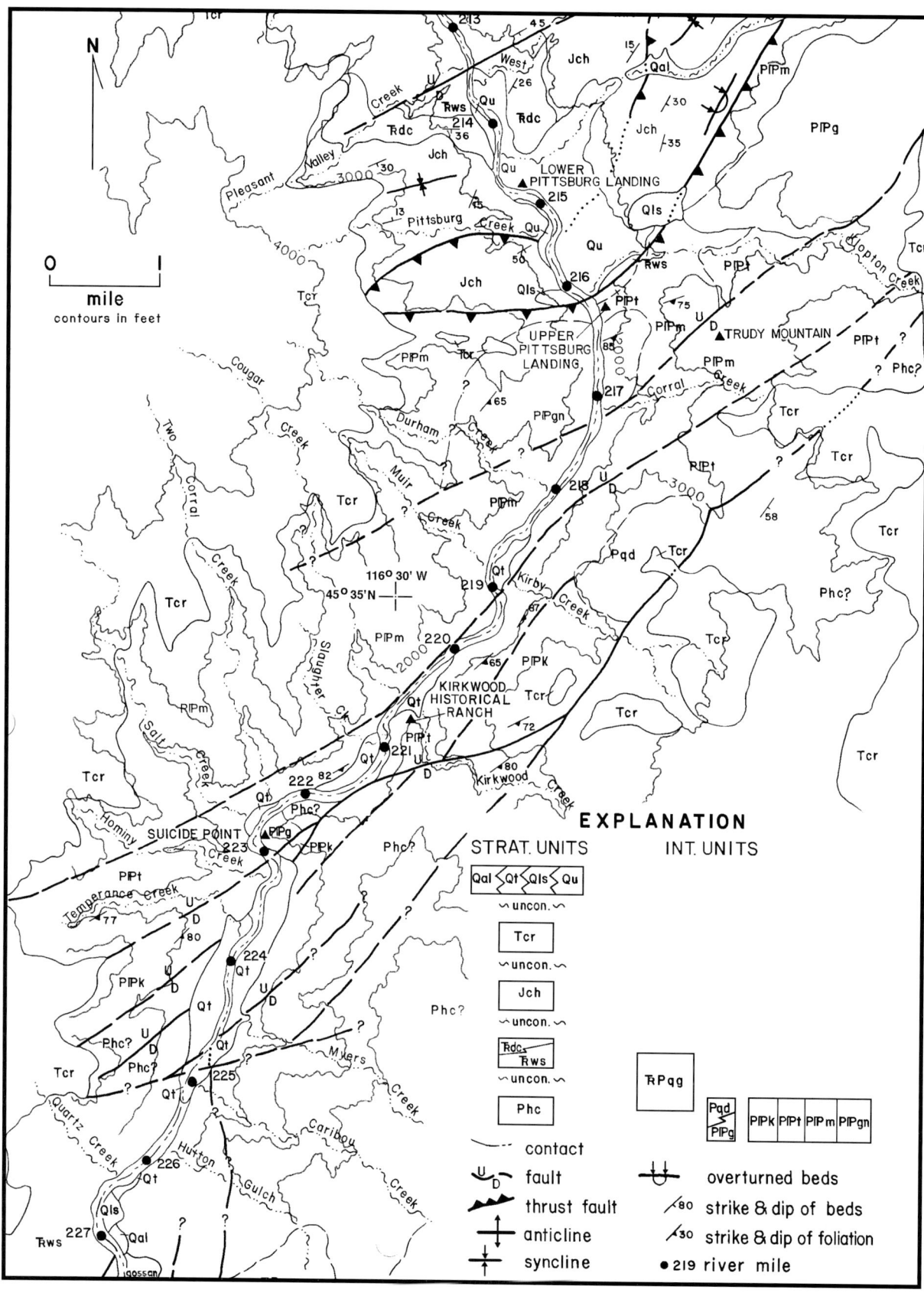

Metamorphism, based on $^{40}Ar/^{39}Ar$ radiometric data that were derived from amphibole crystals, occurred in the age range from 239 to 226 million years with an average of 234 Ma. This age is interpreted as the time when the amphiboles cooled below 500+/-50°C and, therefore, represents an upper age limit on the peak of metamorphism. The mafic gneiss and amphibolite, however, were intruded by at least two undeformed and essentially unmetamorphosed plutons that have U/Pb magmatic crystallization ages of 256 and 246 Ma. These radiometric dates imply that both the initial cooling age(s) of the dikes and of the undeformed plutons and the age of metamorphism are not well understood. For example, was the Middle and Late Triassic metamorphic event related mostly to heat generated by movement along the faults? Or was the heat caused by crystallizing magmas at depth? Certainly, much more geologic work is needed in the Cougar Creek Complex before it is fully understood.

A **silicic** protomylonite from a dike unit that is exposed at Corral Creek (mile 217) yielded a U/Pb radiometric age of about 262 million years. Nicholas Walker (see annotated bibliography) also determined a somewhat questionable U/Pb radiometric age of 309 million years on zircons from a quartzofeldspathic gneiss layer. This is the oldest age yet determined from the complex and the oldest from the Wallowa terrane. If this age is confirmed after additional radiometric age determinations, then we can conclude that the arc initially fired up in the Early Pennsylvanian period. The Wallowa terrane, therefore, may be at least that old, and I presume that it may even be as old as 350 million years, the estimated age of the oldest known sedimentary rocks from the adjacent Baker terrane. Work by Professor Hans Avé Lallemant at Rice University established that extensive left-lateral strike-slip movement occurred within this complex in the Late Triassic.

At mile 223, across from Hominy Creek, Suicide Point looms ominously above the river. The name "Suicide Point" was given to this large piece of real estate because of the steep cliffs bordering the trail that cuts around it. This imposing rock mass influenced the river course, which makes successive 90° turns. A fault zone separates the sheared Permian (?) strata from the Cougar Creek Complex east of Suicide Point (Figure 58).

Stop at Salt Creek (mile 222.5) in Oregon, cross the creek, and walk south on the trail for less than half a mile where a Columbia River Basalt dike cuts through rocks of the Cougar Creek Complex. The hot basalt magma,

Figure 57. Geologic map of Segment 4, which includes several units within the Cougar Creek Complex. See Figure 38 for a complete explanation of all rock and sediment units in Hells Canyon.
Geologic mapping completed by the author.

FIGURE 58. Fault zone separating Permian(?) strata (on the right) from the Cougar Creek Complex.

Figure 59. Xenolith of gabbro from the Cougar Creek Complex in a dike of basalt (Columbia River Basalt Group) along the trail south of Salt Creek.

during its journey upward, forcefully dislodged fragments (xenoliths) of the Cougar Creek Complex and incorporated them (Figure 59). The xenoliths are composed of gabbro and quartz diorite.

The sand beach at Salt Creek is rapidly changing. Sand dunes and an ancient river bar, both of which occur above present river level, are being eroded by the sediment-starved river. Before the dams were built, periodic large floods would bring sediment-clogged water down the river. At Salt Creek, where the river velocity was slowed after shooting through the upstream narrows near Suicide Point, the sediment would settle out, thereby building a high sand bar and beach. Winds picked up loose sand and silt from the beach and deposited them nearby, forming dunes. Now, floodwaters are unable to replenish the sand.

At mile 221, just south of Kirkwood Creek, a nearly vertical dike zone more than a quarter mile wide boldly crops out on the Idaho side of the river (Figure 60). These dikes are metamorphosed gabbro, diorite, quartz diorite, rhyolite, and basalt (many now are mylonite, gneissic mylonite, and amphibolite). Stop just around the bend at Kirkwood Creek (mile 221.5) and walk back to the dike zone. Alternatively, walk to the museum and then north across Kirkwood Creek; about 100 yards north of that creek one can observe a similar dike zone (Figure 61).

Directly across the Snake River from Kirkwood Creek is the mouth of Cougar Creek. A wooden bridge crosses Cougar Creek. This steep tributary canyon is the type locality for the Cougar Creek Complex. I named the Cougar Creek Complex in 1968 after Dave White and I had scratched our heads over the significance of this rock unit in the geologic history of Hells Canyon.

A U.S. FOREST *Service double-cabin hugs the north wall of Salt Creek a quarter mile from the beach. The Forest Service bought the cabin from Lem Wilson when the area became part of the Hells Canyon National Recreation Area (HCNRA). Because of constant downsizing and budget decreases, they have difficulty in keeping this and other structures maintained. The Forest Service and other Federal (and State) agencies can use volunteer help to maintain facilities and trails. Besides help with repairs and maintenance, they also can use assistance with interpretive signs, brochures, and wildlife and plant inventories. Stop in and tell them that you can help. Volunteer.*

VISIT THE KIRKWOOD *Historical Ranch to better appreciate the canyon's late history. From 1933 to 1944, the Jordans lived at the Kirkwood Creek Ranch, including Len Jordan who later became governor of Idaho and then a U.S. senator. Grace Jordan, in 1954, published a book titled "Home Below Hells Canyon" which tells the story of the Jordan family's tenure here. Native American house pits occur along the north side of the Kirkwood Creek road about one half of a mile above the ranch and a thick pile of Mazama Ash crops out behind the Carter cabin. (Watch out for poison ivy.) You can read the history of this area in Carrey, Conley, and Barton (1979, p. 249-264).*

My first visit to the ranch was in 1968. It was then owned by Bud Wilson. Dave White and I approached a ranch hand and asked if we could pitch our tents on the fan below the bunk house (now the museum). Instead, he offered us the bunk house. He fed us beans, mystery meat tainted with the smell of venison, biscuits, and cowboy coffee. We bedded down for the night on well-worn mattresses. An octagon-barreled 30/30 Winchester graced the door of the inner room and a dirty Stetson hat was draped over a nail in the corner. The next morning I asked the hand if I could buy the rifle. "Naw," came the reply. "It belongs to one of the ranch hands who left it here last Fall. But" he continued, "you can have that old hat—no one has claimed it for a decade or so." I accepted the hat, had it cleaned and blocked that Fall, and wore it to classes at Indiana State University where I taught. I wore it out while I worked in the canyon during the next several years.

The canyon slices lower and lower into the Cougar Creek Complex north of Kirkwood Creek. The mixed silicic and mafic dikes observed near Kirkwood Creek are replaced progressively by a zone (miles 218-217) composed almost entirely of mafic dikes (Figure 62) and then by a zone of am-

phibolite, gneissic gabbro, and gabbro near Upper Pittsburg Landing. These mafic igneous rocks are the oldest rocks of the Cougar Creek Complex and are informally called the Trudy Mountain unit because of the excellent exposures on Trudy Mountain (Figure 57). A hike along the river trail between miles 217 and 216 will cross most rock types in the Trudy Mountain unit.

An undeformed quartz diorite pluton of Permian age crops out along the ridge above Kirby Creek. The absence of significant deformation and metamorphism (primary biotite and hornblende are present) in this pluton suggests that some deformation occurred in the dikes of the Trudy Mountain unit before about 246 Ma (U/Pb radiometric age of this pluton). These field and age relationships provide additional evidence that a large part of the Cougar Creek Complex was deformed before the Triassic. In fact, I suspect that most rocks in the Cougar Creek Complex crystallized in the Permian, but that some may be older than Permian.

SEGMENT 5 [UPPER PITTSBURG LANDING TO WEST CREEK—MILES 216 (31) TO 213.5 (33.5)]: THE PITTSBURG LANDING AREA

The river traverses a wide valley, the Pittsburg landing area, to its confluence with West Creek (Figure 63). A road about sixteen miles long heads east from Lower Pittsburg Landing and ends near Whitebird, Idaho.

Pittsburg Landing has more geologic diversity than any other place in Hells Canyon. Dave White and I walked into this valley in 1968. We were greatly impressed by the geology. In fact, Dave was so enthralled by the challenge of these rocks that he came back after a tough tour of duty in Viet Nam and mapped the area for a master's degree.

Almost the entire blueprint of Pittsburg Landing geology can be seen from the top of hills just east of Upper Pittsburg Landing (Figures 64 and 65). A short hike puts you on a water-cut rock terrace that developed during the Bonneville Flood. This rock terrace, nearly 600 feet above river level, is now a **wind gap** and marks the flood's maximum erosive height. Be sure to observe the rounded boulders on the way to the top; they were brought there by the Bonneville Flood.

Figure 60. Dike zone south of Kirkwood Creek in the Cougar Creek Complex.

Figure 61. Similar dike zone to that shown in Figure 60. These rocks crop out along the east side of the terrace just north of Kirkwood Creek.

Figure 62. Mafic dikes (now amphibolite) in the Trudy Mountain unit of the Cougar Creek Complex. Notice the three relative ages of dikes as marked by offsets *(Color section, page 94)*.

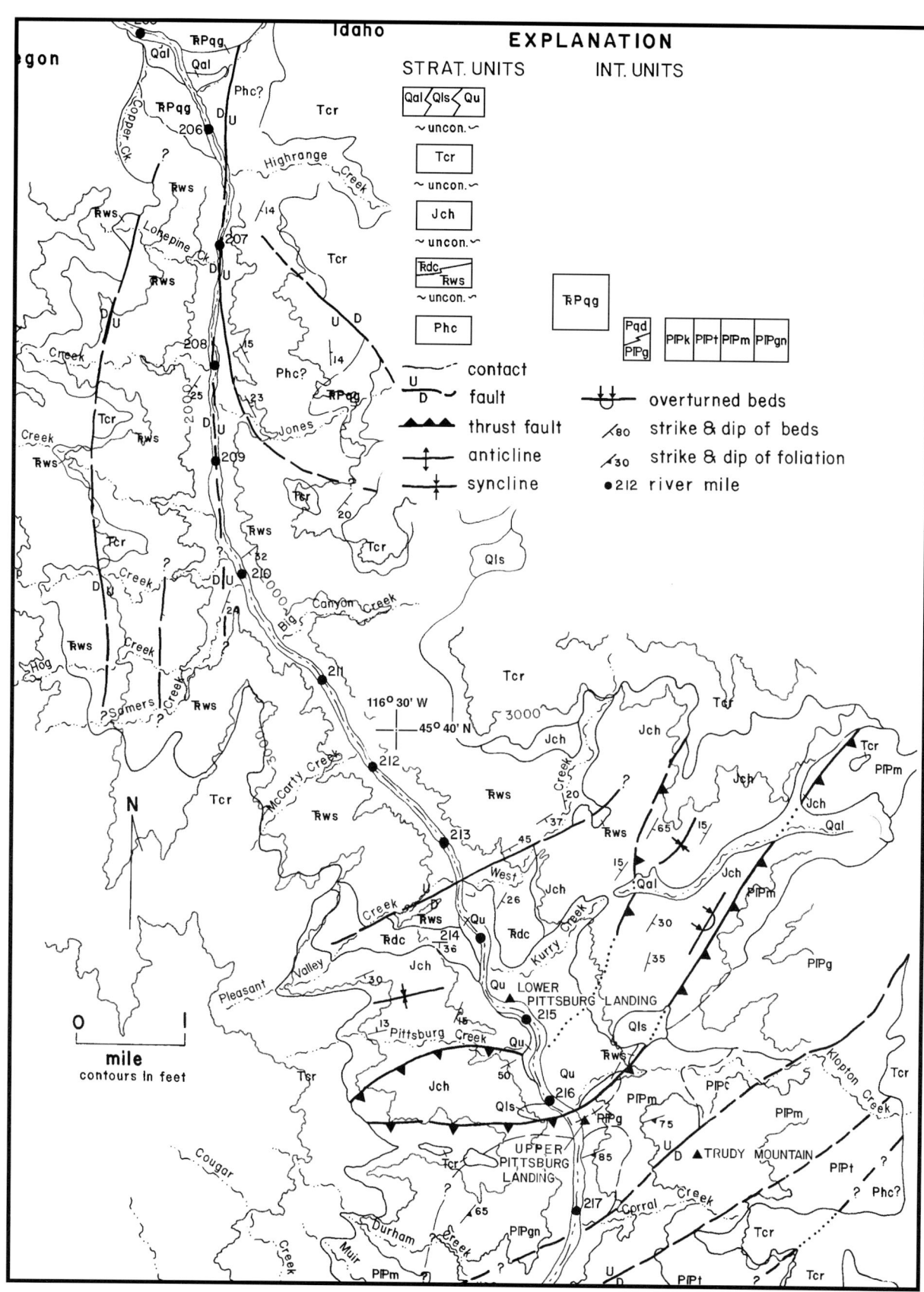

The stratigraphic column at Pittsburg Landing (Figure 66) has several units. Triassic rocks, informal units within the larger stratigraphic framework of Hells Canyon, consist of the Middle Triassic Big Canyon unit of the Wild Sheep Creek Formation and the Upper Triassic Kurry unit of the Doyle Creek Formation. These units are quite different from most other parts of those Triassic formations. For example, the Big Canyon unit consists primarily of pillow basalt breccia and pillow basalt lava flows (Figure 67). The thickness of the pillowed sequence is more than 1,000 feet. Interfingered with the breccias and flows are small stringers of limestone, abundant sandstone, and tuff. The rocks in the Big Canyon unit are more weakly metamorphosed than rocks in most other parts of the Wild Sheep Creek Formation, but are still in the greenschist facies.

The Kurry unit of the Doyle Creek Formation bears no close similarity to the rest of that formation and the correlation was only assigned because of age and its stratigraphic relationship with the Big Canyon unit. The Kurry unit rocks (Figure 68) are mostly sandstone, tuff, siltstone, and silty limestone, whereas the Doyle Creek Formation rocks in other parts of Hells Canyon are predominantly maroon coarse-grained pyroclastic and epiclastic deposits.

PITTSBURG LANDING *boasts the only oil well drilling attempt in Hells Canyon. In the early 1920s, George Wood drilled a 500 foot well. The Jurassic rocks have a high organic carbon content and, if conditions were right, oil may have been generated. However, the right conditions just didn't exist. The well was drilled near Upper Pittsburg Landing, but I have not seen the well head. The Circle C ranch held title to the land in the Pittsburg landing area from the early 1930s until it was acquired by the U. S. Forest Service as part of the Hells Canyon National Recreation Area in the early 1970s. Read more about the Pittsburg Landing area in Carrey, Conley, and Barton (1979).*

Figure 63. Geologic map of Segment 5, the Pittsburg Landing area.

Geologic mapping completed by David White, James D. L. White, Jim O'Connor, and the author.

Figure 64. The Klopton Creek thrust fault separates the Cougar Creek Complex from the Coon Hollow Formation northeast of Upper Pittsburg Landing.

Figure 65. West side (mostly Oregon) of the Pittsburg Landing area. Note the major geologic units and the high Bonneville Flood terrace.

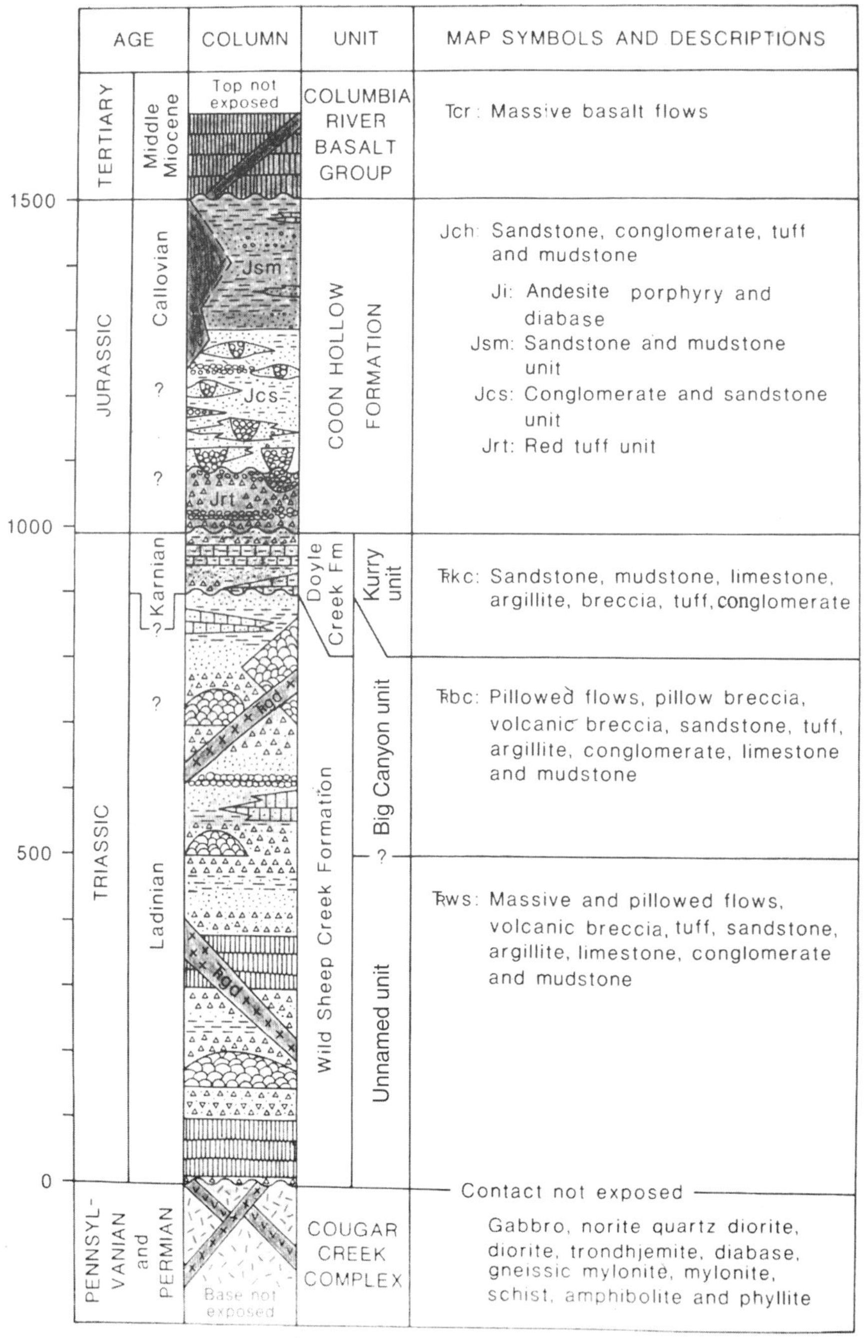

74 ISLANDS AND RAPIDS

AGE	COLUMN	UNIT	MAP SYMBOLS AND DESCRIPTIONS
TERTIARY / Middle Miocene	Top not exposed	COLUMBIA RIVER BASALT GROUP	Tcr: Massive basalt flows
JURASSIC / Callovian / ? / ? / ?	Jsm / Jcs / Jrt	COON HOLLOW FORMATION	Jch: Sandstone, conglomerate, tuff and mudstone Ji: Andesite porphyry and diabase Jsm: Sandstone and mudstone unit Jcs: Conglomerate and sandstone unit Jrt: Red tuff unit
TRIASSIC / ?Karnian / ? / Ladinian		Doyle Creek Fm / Kurry unit	℞kc: Sandstone, mudstone, limestone, argillite, breccia, tuff, conglomerate
		Big Canyon unit / Wild Sheep Creek Formation	℞bc: Pillowed flows, pillow breccia, volcanic breccia, sandstone, tuff, argillite, conglomerate, limestone and mudstone
		Unnamed unit	℞ws: Massive and pillowed flows, volcanic breccia, tuff, sandstone, argillite, limestone, conglomerate and mudstone
PENNSYL-VANIAN and PERMIAN	Base not exposed	COUGAR CREEK COMPLEX	Contact not exposed Gabbro, norite quartz diorite, diorite, trondhjemite, diabase, gneissic mylonite, mylonite, schist, amphibolite and phyllite

Figure 66. Stratigraphic column at Pittsburg Landing. The Big Canyon and Kurry units are shown respectively as the Wild Sheep Creek and Doyle Creek formations on the geologic map of this segment and the Coon Hollow Formation is not divided into the units shown in this column. Left margin represents thickness above Cougar Creek Complex (in meters).

Figure 67. Pillow basalt lava flow in the Wild Sheep Creek Formation near the pass above Upper Cannon Lake, Seven Devils Mountains. The pillow lavas are similar to many of the flows in the Big Canyon unit. These rocks formed along the flanks of an underwater volcano as lava was being erupted into the water.

I suspect that after additional mapping is completed the Kurry unit will be given either a new formation name or else assigned to the upper part of the Wild Sheep Creek Formation.

The Jurassic rocks are of particular interest in the Pittsburg Landing area because of their unique geologic setting and abundant fossils. In this small area, the rocks change in depositional environment, bottom to top, from volcanic (tuffaceous) to fluvial (river and streams) to shallow-water marine and then to deep-water marine. These Jurassic rocks range in age from approximately 180 to 165 million years old (Bajocian to Callovian) and were deposited upon deeply eroded Permian and Triassic rocks. Our interpretations show that a local basin formed on what is now the northern part of the Wallowa terrane; the terrane gradually subsided until finally, by Late Jurassic time, it was covered by ocean water.

The Klopton Creek thrust fault (Figure 69) separates the rugged, and topographically high, metamorphic rocks of the Cougar Creek Complex on the south from deformed Triassic and Jurassic sedimentary rocks on the north. On the Oregon side of the river, the thrust fault overlies the youngest unit (turbidite unit) of the Coon Hollow Formation. This turbidite unit is bounded on the south by the Klopton Creek thrust fault and on the north by an unnamed thrust fault that roughly parallels Backpasture Gulch and Pittsburg Creek from southwest to northeast. Small fault slivers of Triassic rocks crop out near the Klopton Creek thrust fault in lower Klopton Creek.

The deep-water turbidite unit has a wedge-shaped map pattern and consists mostly of calcareous sandstone and siltstone in graded beds. Many of the sandstones consist primarily of limestone fragments. Many others have clasts of volcanic rocks in matrices composed entirely of calcite. The sandstone beds were deposited by successive turbidite flows as the basin deepened. Some of the finer grained mudstones have fossil ammonite molds that indicate a Callovian (late Middle Jurassic) age. A relatively young landslide, formed in these relatively weak sediments, has moved part way down the wall of the canyon (Figure 69).

Spectacular terraces of Bonneville Flood origin and an alluvial fan are apparent on the Idaho side at mile 216. Farther to the east, remnants of a large landslide can be observed not far below the skyline (Figure 70). This

Figure 68. Outcrops of sandstone and siltstone of the Kurry unit. Many of the sedimentary rocks are tuffaceous.

Figure 69. The Klopton Creek thrust fault. Gneissic metamorphic rocks and gabbro are faulted over the turbidite unit of the Jurassic Coon Hollow Formation at this location. Farther east the thrust fault is in contact with the conglomerate and sandstone unit of the Coon Hollow Formation. Note the small landslide just north of the fault that formed in highly fractured sandstone and siltstone of the Jurassic turbidite unit.

Figure 70. Ancient landslide at Pittsburg Landing. The landslide occurred prior to the Bonneville Flood.

Figure 71. Petroglyphs are inscribed on large gabbro boulders at Pittsburg Landing. The boulders were carried by a landslide (Figure 70) to the site and subsequently rounded and fluted by sediment-charged floodwaters.

Figure 72. Delicate fossil fern in the Jurassic Coon Hollow Formation *(Chapter 3 frontispiece, page 44)*.

landslide occurred before the Bonneville Flood because the landslide debris was reworked and smoothed by the flood. The large rocks at the petroglyph site (Figure 71), near mile 215 on the Idaho side, are composed almost entirely of gabbro and were transported by that landslide from the steep hill in the distance. Rapidly flowing and debris-rich water of the Bonneville Flood (and possibly other floods) fluted and rounded the boulders. These rounded and fluted surfaces must have intrigued the Native Americans who responded by chiseling figures on the rocks. Some of the petroglyphs resemble animals and weapons, but archeologists are not in agreement about the meanings of their messages—if any were intended. The site, however, is an important fragment of messages from the past.

The thrust fault between the turbidite unit and the conglomerate and sandstone unit of the Coon Hollow Formation occurs at mile 215 near the mouth of Pittsburg Creek. Generally, thrust faults place older rocks above younger rocks. In this case, however, younger rocks are faulted over older rocks. Apparently, the turbidite unit of the Coon Hollow Formation, which is stratigraphically younger than the conglomerate and sandstone unit, was tectonically detached and subsequently pushed along a relatively flat thrust fault over the more competent conglomerate and sandstone beds. The turbidite unit is isoclinally folded. A deep cleft in the conglomerate near the mouth of Pittsburg Creek outlines a deeply eroded dike of Columbia River Basalt.

Several features are evident from above the launch site and small visitor's center at Lower Pittsburg Landing. A short stream, fed by a spring, enters the river near the launch site. The source for this stream is water from Kurry Creek. During the summer dry season, Kurry Creek water disappears at the upper part of the terrace near the ranch house, and from there the water probably follows an old channel deposit under the Bonneville Flood terrace to its outlet near the launch site. The camp sites and parking spaces are on the Kurry Creek alluvial fan. Above the fan to the east rises a Bonneville Flood terrace. Just east of the launch site near the river are outcrops of mudstone of the Coon Hollow Formation. Plant fossils, mostly delicate ferns, were recovered from those outcrops (Figures 72 and 73). Across the river in Oregon, just slightly northwest of the launch site, conglomerate of the Jurassic Coon Hollow Formation fills an ancient channel that had been cut into the underlying Triassic Kurry unit (Figure 74).

Mazama Ash is well exposed along the alluvial fan on the Oregon side of the river at mile 214 (Figure 75). Directly above the fan, both the Triassic Kurry unit and the conglomerate and sandstone unit of the Coon Hollow Formation are well exposed. A hike up the deepest arroyo will take you through fossil-bearing (ammonites and flat clams) Kurry unit rocks; farther up that hill, fossils in the Coon Hollow Formation include ferns, grasses, and wood fragments.

Bonneville Flood terraces line the Idaho side of the river between Kurry and West creeks. To the east along the lower skyline in Idaho, dipping strata of the Coon Hollow Formation can be seen. The mouth of West Creek (mile 213.5) marks the northern end of the Pittsburg Landing area. About 300 feet above the mouth of Pleasant Valley Creek, to the west in Oregon, gray-colored gravels are apparent. These gravels were deposited by the Bonneville Flood, probably under a backeddy current that was caused by the narrow stricture at the north end of Pittsburg Landing. The terrace high above the river in Oregon is littered with boulders dropped there by the Bonneville Flood.

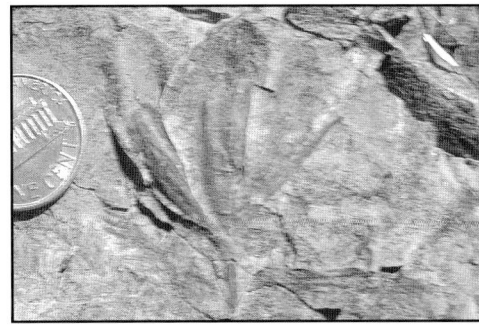

Figure 73. Fossil *Gingko* leaves that were collected from the conglomerate and sandstone unit in the Jurassic Coon Hollow Formation.

Figure 74. A deep stream channel cut into the Triassic Kurry unit crops out on the Oregon side of Pittsburg Landing between Pittsburg and Pleasant Valley creeks. The stream channel was subsequently filled with conglomerate and sandstone of the Jurassic Coon Hollow Formation.

Figure 75. Mazama Ash exposed at Pittsburg Landing in an alluvial fan on the Oregon side, across from the mouth of West Creek.

PLEASANT VALLEY *Dam was slated to be built at the north end of Pittsburg Landing (mile 213.5). The concrete monolith would have risen to a height of 535 feet with impounded water reaching 33 miles upstream, nearly to the base of Hells Canyon Dam. Faded yellow paint marks the height of the intended dam. The proposal for building additional dams in the canyon sparked a twenty year struggle between different factions that culminated in 1975 when the Hells Canyon National Recreation Area bill was passed by both the Senate and the House. For more information, read Carrey, Conley, and Barton (1979, p. 284-287).*

Figure 76. Geologic map of Segment 6. Geologic mapping completed by the author and David White.

Figure 77. Massive beds of pillow basalt lava flows and flow breccias in the Wild Sheep Creek Formation (Big Canyon unit) north of Pittsburg Landing.

SEGMENT 6 [WEST CREEK TO GETTA CREEK— MILES 213.5 (33.5) TO 205.5 (41.5)]

The canyon narrows dramatically at mile 213.5. Although less spectacular than the landscape in Segments 1 and 2 below the Hells Canyon Dam, the canyon walls are nevertheless precipitous. Only reconnaissance geologic mapping was completed on this segment of the canyon and more mapping is needed. Dave White made a few traverses in the southern part of the segment, and I spent less than a week looking at rocks in the northern part.

The rocks along the river from south to north are the Big Canyon unit of the Wild Sheep Creek Formation, probable Hunsaker Creek Formation, and gabbro near Copper Creek (Figure 76). The canyon is very straight north of Somers Creek (mile 210) where the river course apparently follows a fault line. Between miles 213 and 208, thick piles of pillow basalt lava flows and pillow basalt breccia crop out (Figure 77). This part of the canyon traverses Triassic rocks that were deposited close to one or more ancient submarine volcanic vents. The river actually cuts through part of an ancient volcano. Pillow-like shapes decorate the outcrops and show on steep cliff faces. At about mile 212 or 211, stop and examine the pillow basalt flows and pillow basalt breccias. Most outcrops are composed of breccia, but some are pillowed flows.

The southern part of this river segment has an abundance of shiny rocks near river level. Most are merely the result of river polishing; others, however, have thin coatings of iron and manganese. Relatively harder pillows and pillow fragments are more prone to polish than the supporting **matrix** and in places resemble bowling balls.

There are few maintained trails and no major roads along the river between Lower Pittsburg Landing and the mouth of the Grande Ronde River, although ranch access roads and cattle trails do parallel the river in places. A trail parallels the Oregon side of the river south of the Copper Creek resort and then climbs to the top of the bench and from there descends to Pittsburg Landing. Side hikes require more care and preparation, but that shouldn't deter the interested hiker from walking cross country to enjoy some of the flora, fauna, and rocks. Within this segment, I recommend a one-day hike up

Hells Canyon Dam to Grande Ronde River 79

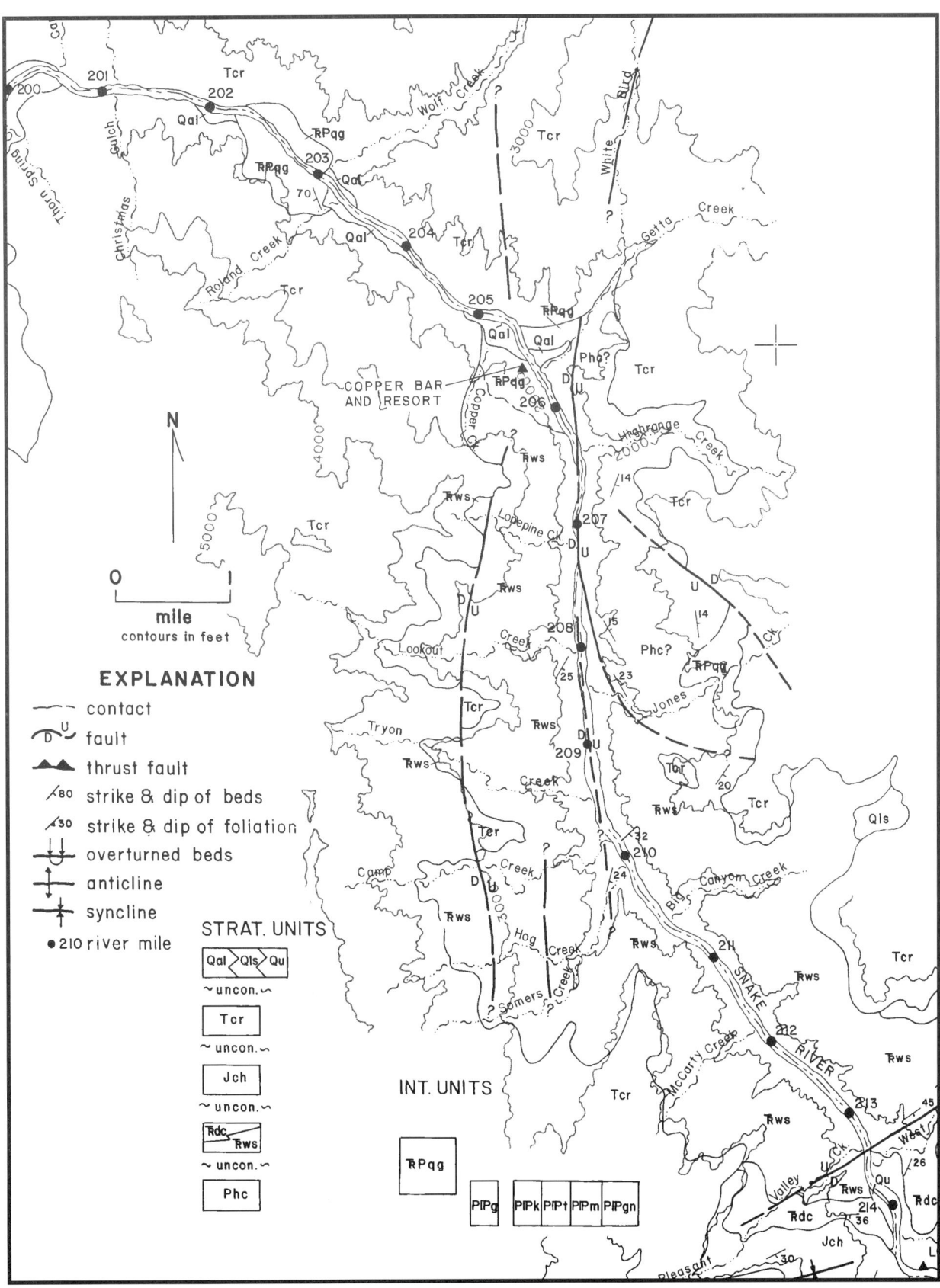

Big Canyon Creek to see the vacated, and now dilapidated, ranch buildings near the headwaters of the creek. The fields and building are on a large slump of the Columbia River Basalt Group (Qls on the map). Slumping of basalt like this is a remarkable method of widening canyons. If that large rock mass had broken up during downslope movement, the debris would have traveled down Big Canyon Creek all the way to the river. Another temporary dam would have formed similar to the one that formed at Rush Creek. And there would be another spectacular rapids in Hells Canyon at the mouth of Big Canyon Creek.

In Idaho, across the river from the mouth of Lone Pine Creek, rocks change characteristics and are mostly light-green sandstone and siltstone, in contrast to the coarse volcanic breccias and flows of the Wild Sheep Creek Formation. A fault separates the units, but mapping has not yet determined the extent of that fault. Some of the rocks are turbidite sandstone beds (Figure 78), with graded bedding from coarse to fine. Most of the lower reaches of the High Range Creek drainage area are underlain by the same rock unit. Both the lithologies and the stratigraphic succession strongly suggest that the rocks are correlative with the Permian Hunsaker Creek Formation, but no fossils have yet been found.

Gabbro crops out between the mouths of High Range and Getta creeks; it extends across the river into Oregon and rises above Copper Creek. This gabbro is of probable Triassic age and may be part of the same group of Triassic plutonic bodies that crop out further downstream and are described in the next segment. Rocks in this gabbro were mined by Billy Rankin in the early 1900s. A prospect tunnel is along the trail about a quarter of a mile south of the resort. Rocks in the tailings indicate that copper minerals, if present, are rare.

SEGMENT 7 [GETTA CREEK TO DUG BAR— MILES 205.5 (41.5) TO 196 (51)]

Segment 7 (Figure 79), beginning at Getta Creek in Idaho and Copper Creek in Oregon, cuts through flows of the Miocene Columbia River Basalt Group, plutonic rocks of Triassic (and Permian?) age, and wide flood terraces. Just north of Copper Creek, outstanding exposures of the Columbia River Basalt Group crop out along the river where thick columns of basalt rise like organ pipes. The terrace across the river from the Copper Bar resort was formed by a flood, probably the Bonneville. The high terrace between miles 204 and 203 is also the probable result of the Bonneville Flood.

The river makes an approximate 90° bend that changes from north near the mouth of High Range Creek to west beginning at mile 202, about 5 miles downstream. I don't understand the reason for this bend. It may be related to the superposition of the river from a flow regime that existed before the deep river canyon was cut. The pre-Cenozoic structures change from northeast-trending to northwest-trending in this area and that change may have influenced the late Cenozoic structures which, in turn, may have guided the course of the river.

Gabbro appears from beneath the basalt at the mouth of Wolf Creek and crops out along both river banks downstream for about one mile. This gabbro has a chemical composition very similar to ocean floor tholeiitic basalt and gabbro and is the most "primitive" gabbro thus far studied in the Wallowa terrane. In other words, the gabbro crystallized from magma that had a com-

Figure 78. Turbidite sandstones in an outcrop of the probable Permian Hunsaker Creek Formation along the Snake River near High Range Creek.

Figure 79. Geologic map of Segment 7. See Figure 38 for an explanation of all rocks units and structural symbols.

Geologic mapping completed by the author

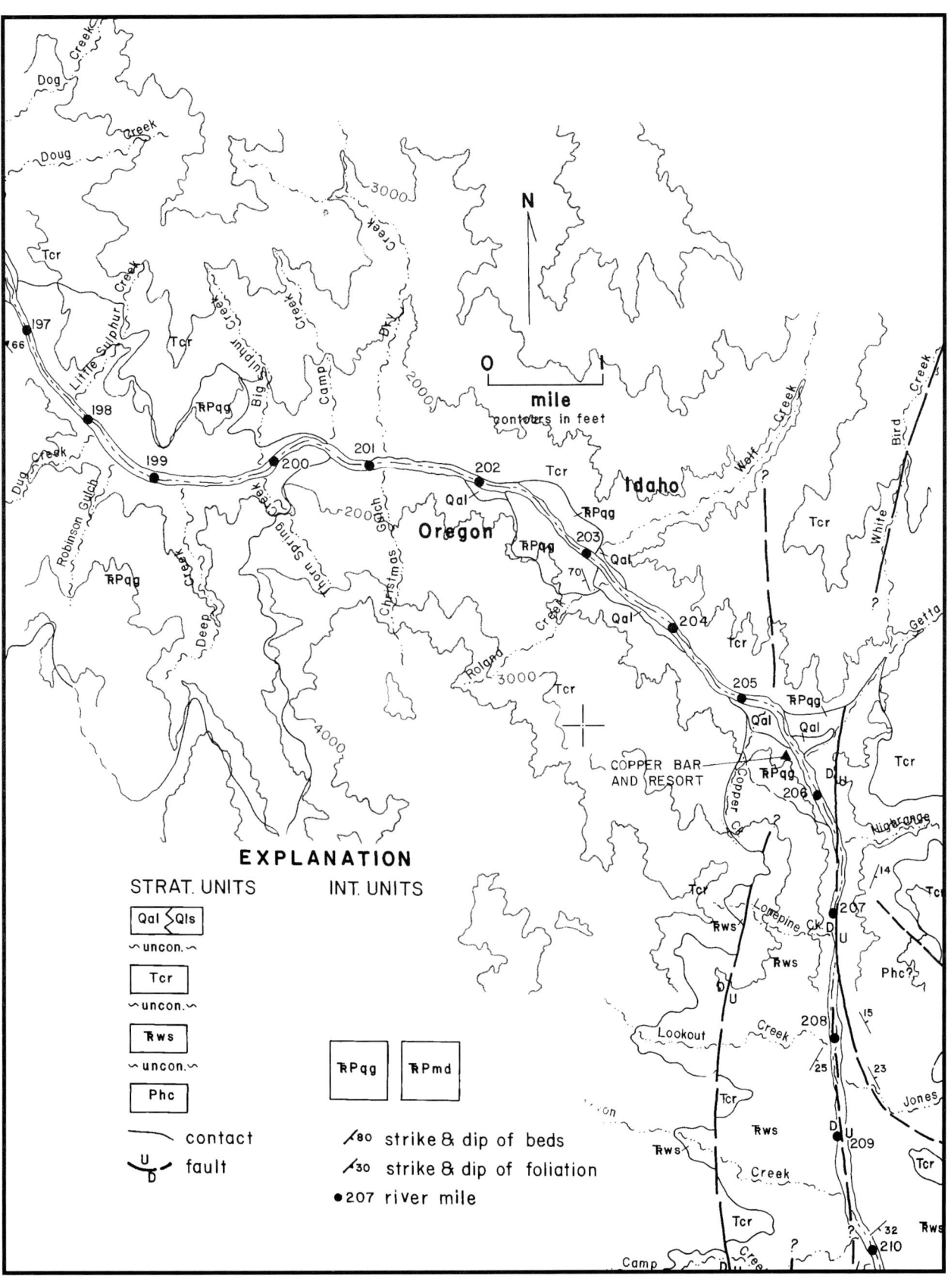

82 ISLANDS AND RAPIDS

A TRAGEDY occurred at Deep Creek (mile 199) in late May of 1887 when thirty-one Chinese miners were murdered for their gold by a group of outlaws who threw the corpses into the river. Authorities investigated the incident and four men were brought to trial, but insufficient evidence failed to force a conviction. Apparently, the murderers didn't find the gold. The gold content of these gabbroic rocks is low and I suspect that the Chinese miners were placer mining the river and Deep Creek sands. See Carrey, Conley, and Barton (1979) for additional details.

Figure 80. Narrow, feldspar-rich dikes stand out in bold contrast to the dark-colored gabbro host rocks near mile 199. The white band is algae that grew on the rocks during higher water levels and then died.

Figure 81. Gray gravels deposited in a backeddy by the Bonneville Flood as it roared through the Dug Bar area about 14,500 years ago.

position closest to the original mantle melt. Differentiation did not affect the magma as much as it had affected the other gabbro plutons in the canyon region.

Below Five Pine Rapids (mile 202), the river glides through flows of the Columbia River Basalt Group. The Hitchcock Ranch at mile 201 is the last of the true "working" ranches in the canyon. The land on the Idaho side of the river is not part of the Hells Canyon National Recreation Area (HCNRA) and private ownership has been maintained. Another ranch, across the river near Christmas Creek, was abandoned when that area became part of the HCNRA.

Near Dry Creek Rapids (mile 201), plutonic rock bodies again crop out along the river to Dug Bar (mile 196.5; also see segment 8). The rocks types are gabbro, norite, quartz diorite, and trondhjemite. Radiometric ages from zircon crystals (Nicholas Walker's work) indicate that the plutonic rocks crystallized about 231 Ma in the Middle Triassic. The plutons are about the same age as the Wild Sheep Creek Formation and probably are crystallized magma chambers that fed some of the volcanic eruptions. Light-colored dikes cut the darker gabbroic rock bodies between miles 200 and 197 (Figure 80). Some dikes are **pegmatites** that contain large crystals of feldspar and quartz. The dikes are narrower and become more vein-like to the west. A sample from one of these vein-like dikes contains mostly albite and quartz. Plutonic rocks continue to crop out along the river all the way to Dug Bar. Just east of Dug Bar (about mile 196.5) the Columbia River Basalt flows reach river level and border the river on the north side.

Across from the Nez Perce crossing sign in Idaho, Bonneville gravels hug the canyon walls of a small tributary Creek that lies in the Dug Bar area (Figure 81). Dark brown and black lava flows of the Columbia River Basalt Group are decorated with these light gray gravels. The terraces at Dug Bar probably formed during the Bonneville flood.

SEGMENT 8 [DUG BAR TO COTTONWOOD CREEK—MILE 196 (51) TO MILE 180 (67)]

Segment 8 (Figure 82a,b) transects about 16 miles of the canyon and takes the traveler through rocks of mostly Triassic, Jurassic, and Miocene ages. Bonneville flood features are common. A stratigraphic section (Figure 83) shows that the lava flows are mostly of the Imnaha Formation, the oldest formation of the Columbia River Basalt Group in Hells Canyon.

Between Warm Spring Rapids (mile 195) and Divide Creek Rapids (mile 193), the river parallels a wide zone of sheared rocks that belongs to the Hunsaker Creek and Wild Sheep Creek formations and includes abundant dikes. Mylonitic zones, some of which are dikes, are apparent and, in places, are dominant. Because of the abundance of dikes and the intense shearing, these mylonitic rocks were placed in the plutonic category in the general explanation of Figure 38. However, the reader should recognize that there are also large components of volcanic and sedimentary rocks in this unit.

In contrast to the strong northeast-trending fabric farther south along the canyon, the rock structures and mineral foliations here trend nearly east-west, parallel to the river channel. At about mile 192, near the mouth of the Imnaha River, the river follows a 60° bend and trends north-northwest. Rocks and structures along Segment 8 form the most complicated region in the lower part of the Snake River Canyon, where steep-dipping faults juxtapose very different rock bodies.

CHIEF JOSEPH *and the Nez Perce people crossed the river at the upper end of Dug Bar with their cattle and horses during an attempted escape to Canada. It was the Spring of 1877 and the river was at flood stage. Rest here for a while and listen to the bawling of cattle, the cries of children, the neighing of horses, and the shouting of the young men as they drove the livestock into the brown water. One can imagine the stoic expressions on the faces of Nez Perce elders as they directed the crossing. The Nez Perce people had given up their lands and were confused about the future. A succinct summary of the Nez Perce flight from the Wallowa Mountains to northern Montana is told by Carrey, Conley, and Barton (1979, p. 314-335).*

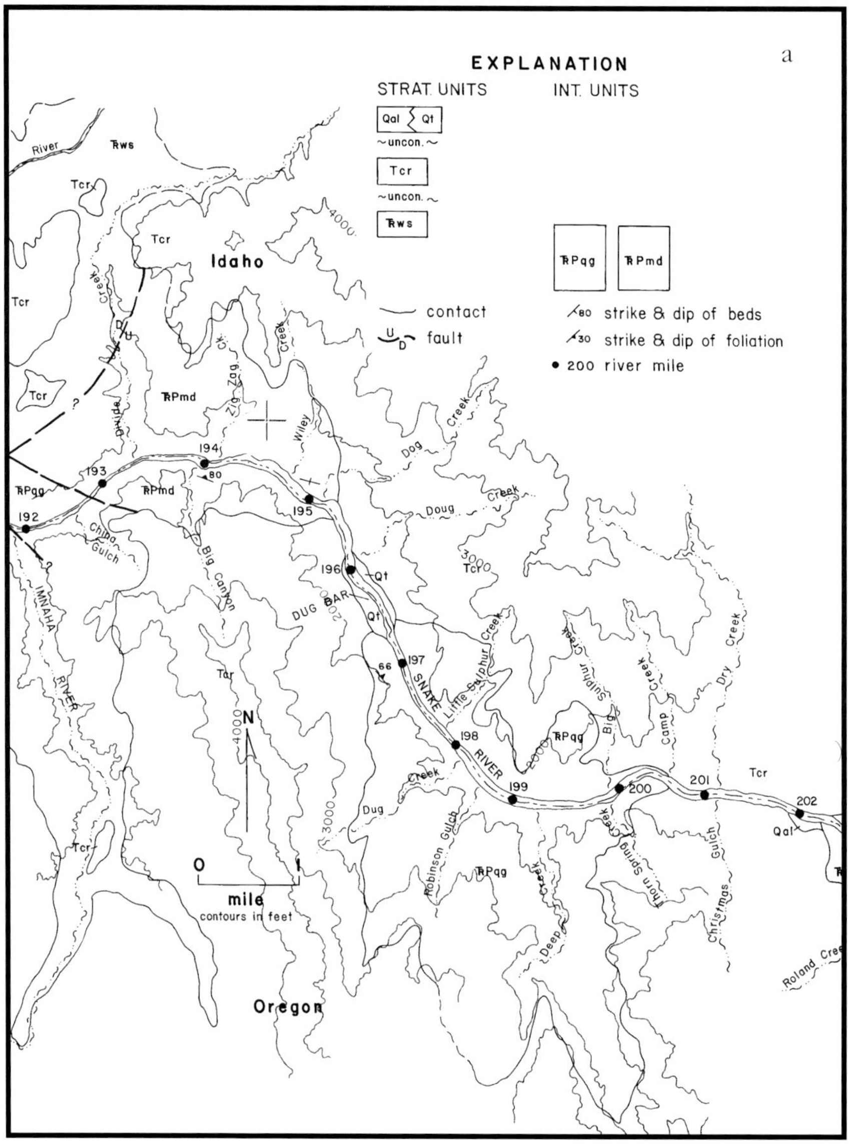

Figure 82a, b. Geologic maps of Segment 8.

Geologic mapping completed by S.-J. Chen, Robert Morrison, Patrick Goldstrand, and the author.

The Imnaha Intrusion, well exposed near the mouth of the Imnaha River, consists of gabbro, norite, diorite, and quartz diorite. A trondhjemitic phase of the pluton has slightly discordant U/Pb radiometric ages of 225 and 228 Ma, indicating that the Imnaha Intrusion is somewhat younger than the Deep Creek body (231 Ma) that parallels the river in Segment 7. All of the larger plutonic bodies in Segments 7 and 8 most likely crystallized as volcano-root plutons beneath Middle and Late Triassic volcanoes.

The most famous of the mines in this part of Hells Canyon is the Mountain Chief Mine (mile 192). The tunnel entrance (Figure 84) lies nearly 100 feet above the Snake River and several hundred feet east of the mouth of the Imnaha River. The mine was tunneled along a fault zone; no other rocks in the immediate area near the fault zone show significant mineralization. Gabbro and diorite that crop out near the east entrance are representative of undeformed parent materials. The small rock dump (tailings) near the west entrance contains some of the minerals that were taken from the ore zone,

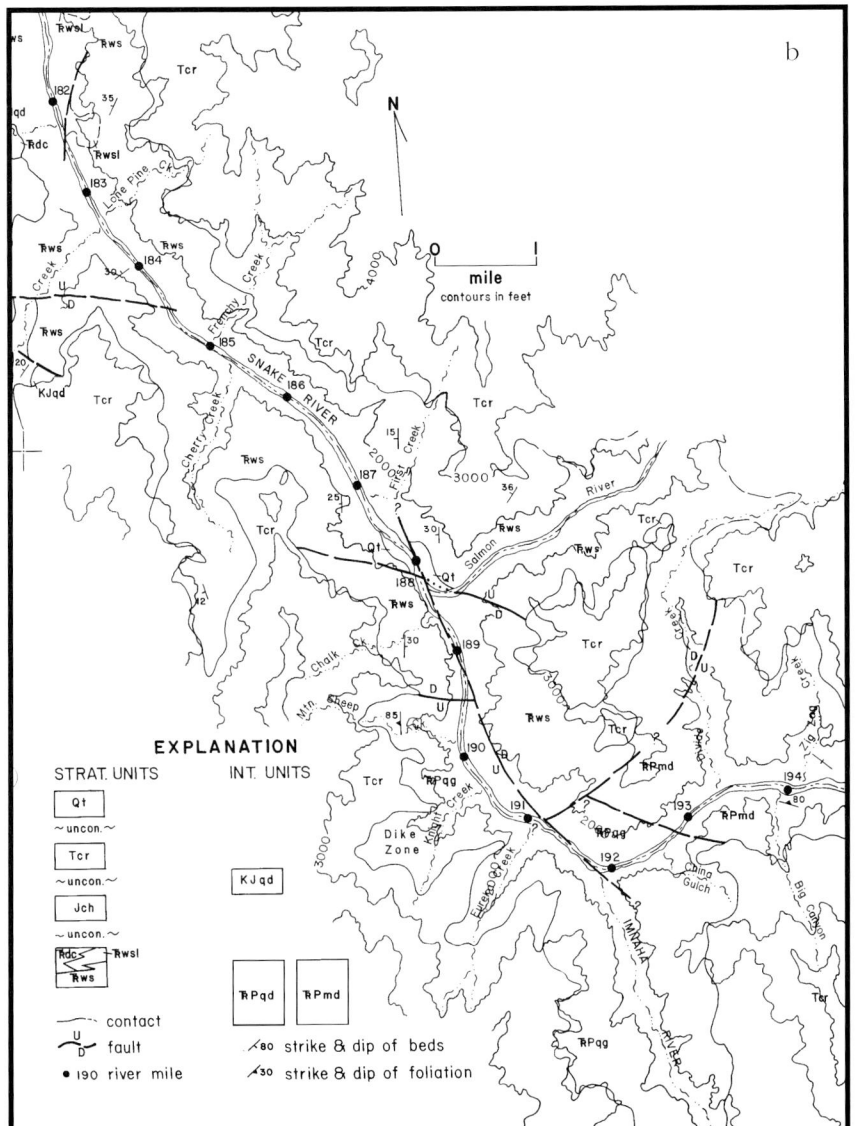

THE REGION near the Imnaha-Snake River confluence experienced a flurry of gold and copper exploration in the late 1800s and early 1900s. Along the canyon walls, prospect pits, mine portals, and ore dumps attest to the fervor of expectant prospectors. The Eureka Mining, Smelter, and Power Company was set up in 1902 to mine and smelt the ore from the area, although the Fargo Company was the primary mining company. The foundation of a mill still stands above Eureka Bar, where a small town was erected with a hotel, grocery store, and post office. The steamboat Imnaha was the workhorse and made thirteen trips from Lewiston without any significant problems. However, in November of 1903, the Imnaha sank, thereby ending the entire enterprise. I encourage you to read more about this venture in Carrey, Conley, and Barton (1979, p. 342-353). Many accounts in western mining history are similar to this enterprise. Only the people who thought up the venture received profits; shareholders ended up with worthless stock.

including shiny black hematite (iron oxide), yellow limonite (iron oxide with water in the mineral lattice), abundant pyrite (fools gold—an iron sulfide), plus some chalcopyrite (copper and iron sulfide), chalcocite (copper sulfide), and malachite (copper carbonate). White, green, and blue minerals that coat many of the outcrops within the tunnel are sulfates.

The Imnaha Intrusion continues northward along the Imnaha Rapids. Abundant pre-Cenozoic dikes, ranging in composition from diabase through rhyolite, form an anastomosing network in the upper reaches of Eureka Creek and Knight Creek drainage basins (Knight Creek enters the Snake River near mile 190.5). These dikes are unique in the canyon and may mark the site of a Middle Triassic volcano; some of the dikes probably were the magma feeders for lava flows that erupted on the underwater flanks of a volcano. Dike compositions include basalt, diabase, andesite, and rhyolite. A fault separates the Imnaha Intrusion on the south from the Wild Sheep Creek Formation on the north near Mountain Sheep Rapids (below mile 190).

86 ISLANDS AND RAPIDS

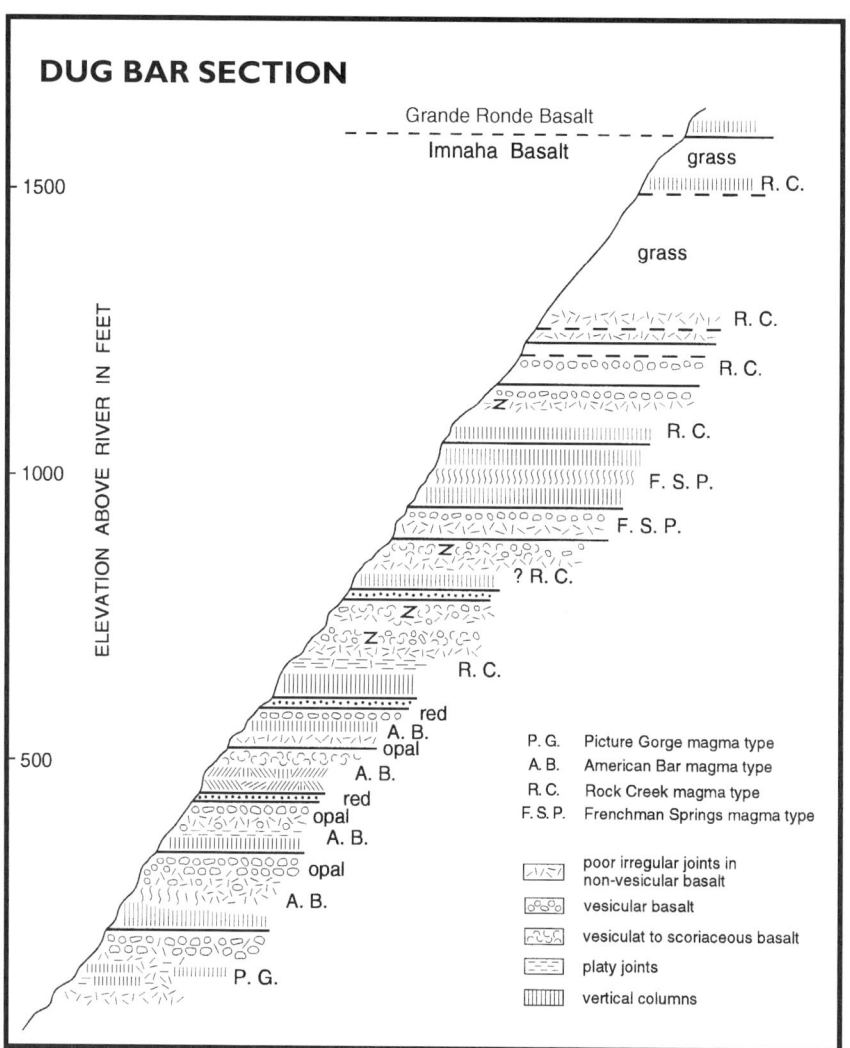

Figure 83. Section of Columbia River Basalt measured by W.D. Kleck, G.S. Holden, and P.R. Hooper. Notice that almost all of the flows belong to the Imnaha Formation, which is the oldest formation of the Columbia River Basalt Group in the region. The z in bold face represents zones where zeolite minerals are abundant. (The section is modified from T. Vallier and P. Hooper, 1976, in a guidebook for a field trip that was organized for the Geological Society of America meeting in Pullman, Washington.)

Figure 84. Entrance to the Mountain Chief Mine. The white area near the bottom of the photo is caused by algae growth during high water.

The narrowest part of Hells Canyon lies between Knight Creek and the mouth of the Salmon River. Iron rings, used for winching steamboats through Mountain Sheep Rapids, are still set in rocks on both sides of the river near the mouth of Mountain Sheep Creek. A similar iron ring, taken from the rocks near Knight Creek, hangs on a wall in the Kirkwood Museum.

The Salmon River joins the Snake River at mile 188.5 (Figure 85). A fault, easily seen from the confluence, breaks the rocks' continuity along the south side of the Salmon River Canyon about one half of a mile above the confluence. The Pullman Mine was tunneled along that fault and mine tailings are still piled near the mine's entrance. Some ore was taken from the mine to Lewiston in the 1920s.

Neither the Salmon River nor any of its tributaries are dammed. From the confluence of the Snake and Salmon rivers to the Salmon River's smallest tributaries, deep in the Idaho wilderness, the river flows unimpeded. Floodwaters with flows greater than 100,000 cfs still boil through the narrow canyon walls of the Salmon River during the spring run off. Dry intervals late in the summer are often marked by a mere trickle of water.

The geology recorded in Hells Canyon is representative of rocks all the way to Slate Creek along the Salmon River Canyon. Stratigraphy, characteristics of plutons, and deformation of the pre-Tertiary rocks are about the same. From near Slate Creek, however, south all the way to the "time-zone" bridge near Riggins, Idaho, the pre-Tertiary rocks are more deformed and metamorphosed. I agree with most other geologists who have worked in the region that the pre-Tertiary rocks (Riggins Group) between Slate Creek and the bridge near Riggins are probably correlative to those of the Seven Devils Group and its associated Permian and Triassic plutons.

At the "time-zone" bridge, an ultramafic body separates very different rocks. I suspect that the rocks south of the bridge are correlative to the Olds Ferry Terrane and those to the north are correlative to the Wallowa Terrane. The ultramafic body probably was squeezed up along a fault zone that formed long before the Blue Mountains Island Arc was accreted to the ancient North American continent. All pre-Tertiary rocks, both along the Salmon River south of Slate Creek and along the Little Salmon River south of Riggins, were deformed during collision of the Blue Mountains Island Arc with the North American continent and subsequent tectonic movements and pluton intrusions related to that suturing event. This dramatic suture zone, named the Salmon River Suture, extends in a north-south direction between McCall, Idaho and Orofino, Idaho.

Relatively high river terraces occur below the confluence of the Snake and Salmon rivers. These terraces are the result of high flood waters from combined Salmon River and Snake River flows and are not the result of the Bonneville Flood. An abundance of shiny rocks lines both sides of the river between miles 188 and 187.

The Wild Sheep Creek Formation is the dominant unit along the Snake River between miles 190 and 180. Pat Goldstrand, who mapped this section of the canyon in 1986, divided the Wild Sheep Creek Formation into five mappable facies. The rocks and general characteristics of the Wild Sheep Creek Formation are different here from those observed farther south. For example, the relative abundance of coarse volcanic breccias and lava flows decreases from south to north and the percentage of siltstone, limestone, and sandstone increases. Siltstone (metamorphosed to argillite) and sandstone beds crop out along the Idaho side of the river in several places (Figure 86);

Figure 85. Hells Canyon near the confluence of the Snake and Salmon rivers. The mouth of Salmon River is on the left. Rocks of the Wild Sheep Creek Formation crop out on all canyon walls near the confluence *(Color section, page 95)*.

Figure 86. Sandstone and siltstone (argillite) beds in the Wild Sheep Creek Formation north of First Creek near mile 187.

Photograph by Pat Goldstrand.

best exposures are upstream about half a mile from the confluence of the Salmon and Snake rivers and just downstream past mile 187 where thin beds of argillite and sandstone dip about 20° to 25° to the northwest. Dikes, probably crystallized feeders for flows of the Wild Sheep Creek Formation, cut the argillite beds and disrupt bedding. Limestone bodies parallel the river and are much more dominant in this part of the Wild Sheep Creek Formation than in outcrops elsewhere. Fossils from the bedded limestone are mostly the flat clam *Halobia* of Karnian (Late Triassic) age. In the southern part of Hells Canyon fossils in the Wild Sheep Creek Formation are predominantly the flat clam *Daonella* of Ladinian (Middle Triassic) age.

In stark contrast to the near absence of beaches upstream along the Snake River above its confluence with the Salmon River, buff-colored sand beaches are abundant between that confluence and the mouth of the Grande Ronde River. A good example of the beaches is at Geneva Bar (mile 184.5). Floodwaters from the Salmon River continue to replenish these beaches with sands that were eroded mostly from the quartz- and feldspar-rich Idaho Batholith. The Idaho Batholith crops out extensively along the Salmon River east of Riggins, Idaho, and is the dominant rock unit in many of the tributary streams. The beaches below the confluence of the Salmon and Snake rivers are similar to those that existed throughout Hells Canyon before the construction of dams.

The mouth of Cherry Creek (mile 185) is very narrow and its water flows out between some large boulders. The creek was named in the late 1800s for the abundance of chokecherry trees that grew in its canyon. Cook Creek (mile 183) is one of the larger creeks to enter the Snake River in this area. Near its mouth abundant shiny rocks are evident. A blowout in early January of 1997 modified the alluvial fan. Just below Cook Creek excellent bedding can be observed in the Wild Sheep Creek Formation. Most of the rocks are volcanic breccia and sandstone.

Notice that limestone beds are common in this section of the canyon (Figure 87). These limestone beds are part of the Wild Sheep Creek Formation and most are older than the Martin Bridge Limestone in Hells Canyon by a few million years. However, rocks from the type locality of the Martin Bridge Limestone along Eagle Creek in the southern Wallowa Mountains are in part late Karnian in age. Are the limestones in the Wild Sheep Creek Formation along this part of Hells Canyon precursors to—or are they equivalent to the lower part of—the Martin Bridge Limestone? The limestone in this part of Hells Canyon contains a wide variety of fossils including crinoids, corals, and clams. The largest limestone body in the Wild Sheep Creek Formation occurs on the Idaho side of the river between miles 182 and 180. It is particularly well exposed near the mouth of Cottonwood Creek where the beds are folded into a plunging **anticline** (Figure 88).

The angular unconformity between the Triassic Wild Sheep Creek Formation and the Jurassic Coon Hollow Formation (Figure 89) is well exposed in this section of the canyon. The unconformity crops out near the river at about mile 179.8 above Cougar Bar Rapids. It can be followed by eye at about mile 180, southwestward into Oregon and southeastward into Idaho. The unconformity separates strata 230 to 225 million years old from strata 160 to 150 million years old. In order to study the unconformity, I suggest that you camp at Lower Cottonwood Creek campground in Idaho, walk about one mile down river to mile 179.8, and then follow the unconformity about one half of a mile to the southeast.

Figure 4

Figure 11

Figure 12

Figure 13

Figure 14

Figure 29

Figure 48

Figure 49

Figure 51

94 ISLANDS AND RAPIDS

Figure 55

Figure 62

Figure 85

96   ISLANDS AND RAPIDS

Figure 114

Figure 121

Figure 87. Limestone near the mouth of Coon Creek.

Figure 88. Folded limestone of the Wild Sheep Creek Formation exposed near the mouth of Cottonwood Creek.

Figure 89. Unconformity (u) between the Wild Sheep Creek Formation (WSC) and the Coon Hollow Formation (Jcs).

Pat Goldstrand Photo/Labels

SEGMENT 9 [COTTONWOOD CREEK TO CHINA GARDEN CREEK—MILE 180 (67) TO MILE 176 (71)]

The rocks change dramatically to sandstone, siltstone, and rare conglomerate of the Upper Jurassic Coon Hollow Formation near mile 180 (Figure 90). This basinal marine sequence is somewhat younger than the youngest Coon Hollow Formation strata exposed at Pittsburg Landing (Late Jurassic Oxfordian fossils here and late Middle Jurassic Callovian fossils at Pittsburg Landing); however, the rocks in both places are part of the same transgressive marine sequence that covered most of the subsiding Wallowa terrane (and most of the entire Blue Mountains Island Arc) in the Middle and Late Jurassic.

The Coon Hollow Formation in this area consists of a lower conglomerate and sandstone unit that is overlain by a finer-grained sandstone and siltstone unit (Figure 91). The lower conglomerate and sandstone unit crops out along the Idaho side of the river at mile 179. Most sandstone and mud-

Figure 90. Geologic map of Segment 9. See Figure 38 for the explanation for all units in Hells Canyon.

Geologic mapping completed by Pat Goldstrand, Robert Morrison, and the author.

Figure 91. Outcrops of the Coon Hollow Formation near the type locality.

Figure 92. Deformed sandstone beds in the Coon Hollow Formation.

stone beds were deposited from turbidity currents in a relatively deep ocean basin. Fragments of ammonites can be found in places within the sandstone beds. Although most clasts in the basal conglomerate were eroded from the Wallowa Terrane, some are clasts of Late Triassic radiolarian chert that were eroded from the Baker Terrane. Such evidence indicates that the Baker and Wallowa terranes were adjacent, in places occurred above sea level as islands, and were being eroded during the deposition of the Coon Hollow conglomerate beds. Beds in the Coon Hollow Formation along this segment of Hells Canyon are greatly deformed by folding and faulting (Figure 92). Stratigraphic continuity can be demonstrated only in small areas. Hornblende-rich diorite sills and dikes cut the unit.

Outcrops of the Coon Hollow Formation continue along both sides of the Snake River to about mile 175.5, just north of the Oregon-Washington border. At that point, the Coon Hollow rocks are cut by an Early Cretaceous (about 135 million years old by K/Ar methods of radiometric dating) quartz diorite pluton. The Oregon-Washington border, at about mile 176, also marks the boundary of the Hells Canyon National Recreation Area.

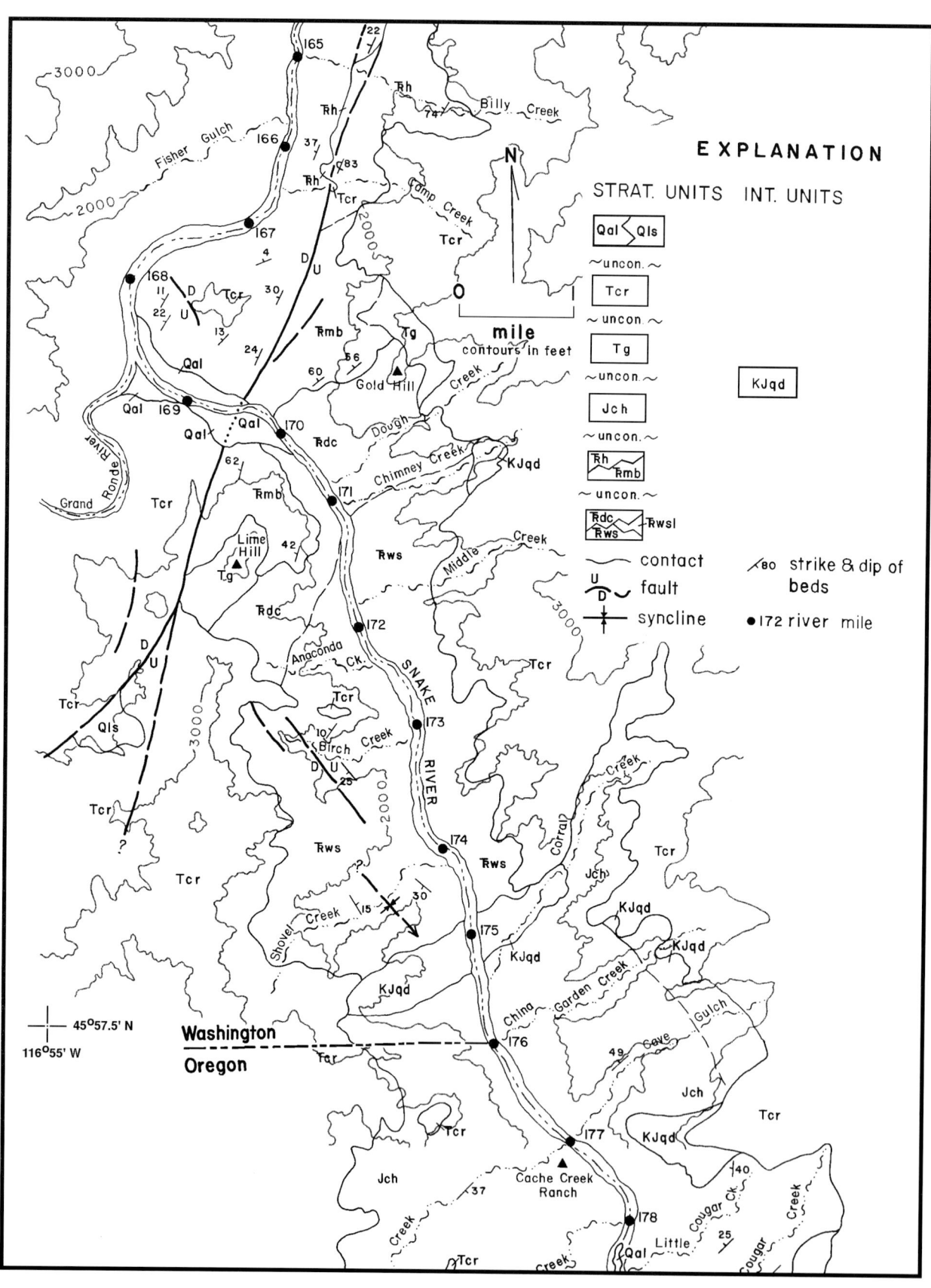

SEGMENT 10 [CHINA GARDEN CREEK TO GRANDE RONDE RIVER—MILES 176 (71) TO 168.5 (78.5)]

Parts of all Triassic strata (Wild Sheep Creek and Doyle Creek formations, Martin Bridge Limestone, and Hurwal Formation) are exposed in this segment of Hells Canyon and so is an Early Cretaceous pluton (Figure 93). Outcrops of the quartz diorite pluton occur on both sides of the river between mile 175.5 and Shovel Creek (mile 174.5). The quartz diorite is not metamorphosed, in sharp contrast to the Triassic plutons that crop out near the mouth of the Imnaha River. At about mile 175, along the riverbank in Idaho, elongate pieces of quartz diorite are stacked like cord wood. Some similar quarried rocks from this locality were used as window ledges and door sills in buildings of Lewis-Clark State College in Lewiston. The rocks were quarried mostly in the early 1900s from outcrops about a quarter mile up Corral Creek. Notice that the plutonic rocks are jointed and form somewhat irregular polygonal outcrops along the slopes of the canyon.

Near Shovel Creek the river leaves the quartz diorite pluton and cuts into strata of the Wild Sheep Creek Formation. These Triassic strata are coarser grained than those farther south in the canyon near the Salmon River; most are breccias and coarse-grained sandstones; some massive basalt flows and dikes occur. A thin band of maroon rocks near mile 170, immediately underlying the Martin Bridge Limestone, correlates with the upper part of the Doyle Creek Formation. To my knowledge, no fossils have been collected from the Wild Sheep Creek and Doyle Creek formations in this segment.

Figure 93. Geologic map of Segment 10.
Geologic mapping completed by S. Reidel, M. Glerup, and the author.

Figure 94. Martin Bridge Limestone in Washington near mile 170. Note the small alluvial fan on the right that formed from one or more debris flows originating about 500 feet above the river. Dark strata beneath the limestone are correlated with the Doyle Creek Formation.

102 ISLANDS AND RAPIDS

Figure 95. Limekiln Fault separating Martin Bridge Limestone (on the right) from Columbia River Basalt Group flows in Idaho near the mouth of the Grande Ronde River. Notice that the basalt flows dip to the west near the fault.

Figure 96. Limekiln fault separating Martin Bridge Limestone (on the left) from the Columbia River Basalt Group in Washington near the mouth of the Grand Ronde River.

Bold outcrops of Martin Bridge Limestone begin at about mile 171. At mile 170, limestone beds crop out on both sides of the river (Figure 94). The limestone is sheared and, to my knowledge, no fossils have been reported. Look north at about mile 169.7 and you will see a more shaley facies of the limestone. This shaley facies and the mudstones farther north are correlated with the Hurwal Formation, a rock unit that is well exposed in the northern Wallowa Mountains.

The Limekiln Fault crosses the Snake River at about mile 169.5 and juxtaposes the Martin Bridge Limestone and Columbia River Basalt Group (Figures 95 and 96). This fault is one of the most spectacular structures in Hells Canyon because of the contrast in color between offset rock units. The displacement along this fault is more than 1,000 feet. If active, the fault could generate a large earthquake (probably a Richter magnitude 5.0 to 6.0) because the length of the fault is greater than 20 miles. To my knowledge, however, there has been no seismic activity along it within recorded history. The fault defines the southeast side of Lewiston basin. Displacement along the Limekiln fault is mostly responsible for the high terrain known as Craig Mountain that forms the skyline southeast of Lewiston, Idaho.

The small settlement of Rogersburg, just south of the Snake River-Grande Ronde River confluence, is built on landforms related to the Bonneville Flood. Bonneville Flood water, previously constricted between canyon walls, spewed out into the wider valley, dropping its sediment load. The Bonneville Flood waters and, later (between about 14,000 years ago and 12,000 years ago), Missoula Flood waters, flowed up the Grande Ronde River for many miles.

Lava flows of the Columbia River Basalt Group line both sides of the river from the mouth of the Grande Ronde River to Lewiston and beyond (Figure 97). Only a few more inliers of pre-Cenozoic rocks lie downstream from the mouth of the Grande Ronde River. One of the inliers occurs at Buffalo Eddy where petroglyphs have been carved into pre-Cenozoic (probably Triassic) rocks.

The river (and road) trip between the mouth of the Grande Ronde River and Lewiston/Clarkston is mostly a trip through lava flows of the Columbia River Basalt Group. Imagine lavas pouring out of fissures and rapidly covering the landscape, in some cases for thousands of square miles. The source of these lavas is the underlying mantle that melted, collected in large magma chambers, and then abruptly emptied out onto the surface. Geologists do not understand the reason for the eruption of this large accumulation of basalt that now covers large parts of eastern and central Washington, western Idaho, and northern Oregon.

A spectacular outcrop of the Columbia River Basalt Group occurs in Idaho across the Snake River from the town of Asotin, Washington. Five- and six-sided columns of basalt create an intriguing mosaic of contrasting geometric forms (Figure 51). The lava flows, part of the Pomona Member, are about twelve million years old. The flows filled arroyos and canyons and cooled to form the bent and tilted columns. Another outstanding outcrop of the Pomona Member lava flows is at Swallow's Nest in Clarkston, Washington, where the flows form the lower part of a Gibraltar-type protuberance into the Snake River Canyon.

The take-out point for float trips and power boats is less than a mile below the mouth of the Grande Ronde River in Washington at Heller Bar. A good road leads from there to Asotin.

Figure 97. Thick pile of lava flows of the Columbia River Basalt Group exposed near the mouth of the Grande Ronde River. These flows belong mostly to the Grande Ronde Basalt.

CHAPTER 4

Geologic Guide Between Oxbow and Hells Canyon Dams

INTRODUCTION

This chapter focuses on rocks that are exposed near and along the Idaho Power Company (IPCO) road that parallels the east side of the Hells Canyon Dam reservoir between the Oxbow and Hells Canyon dams: a distance of about 22 miles. I began my studies here and ultimately mapped this area of Hells Canyon in more detail than in most of the other parts.

I first discuss the Oxbow Dam area. Then a road log begins at the Snake River bridge near the Copperfield campground (mile 0) and ends near Hells Canyon Dam (about mile 22). Geologic maps (Figures 98-101) illustrate the geology of the area; road miles are excluded from the maps, but geographic features are mentioned throughout the text to help orient the reader.

GEOLOGY OF THE OXBOW DAM AREA

The strange configuration of the river channel in the Oxbow region is evident from the air (Figure 102) and from the ground (Figure 103). The configuration, however, is not a river **oxbow** in the classic sense, although it has the appearance of an oxbow that forms along meandering streams. At this oxbow, the river's present-day course follows pre-Cenozoic structural trends. I suspect that as the river cut down from above, it followed zones of weakness that had been imposed into the overlying basalts by older structures.

Plan to study rocks that crop out along the road leading to the Oxbow Dam to gain an understanding of the Oxbow region. To reach that road, assume that you are driving from Halfway, Oregon, to Hells Canyon on Highway 86. Turn right off Highway 86 onto the Baker-Homestead Highway that leads to Brownlee Dam and cross Pine Creek bridge. Turn left at the next intersection (about one quarter of a mile) and onto the road that leads past the power station and then to Oxbow Dam.

From a location near the surge tanks look north across the river to see canyon-filling basalt flows with curved and radiating columnar jointing. Hot lava flowed down an arroyo and subsequently cooled. The shape of the ancient arroyo is evident. These lava flows belong to the Imnaha Formation of the Columbia River Basalt Group.

106 ISLANDS AND RAPIDS

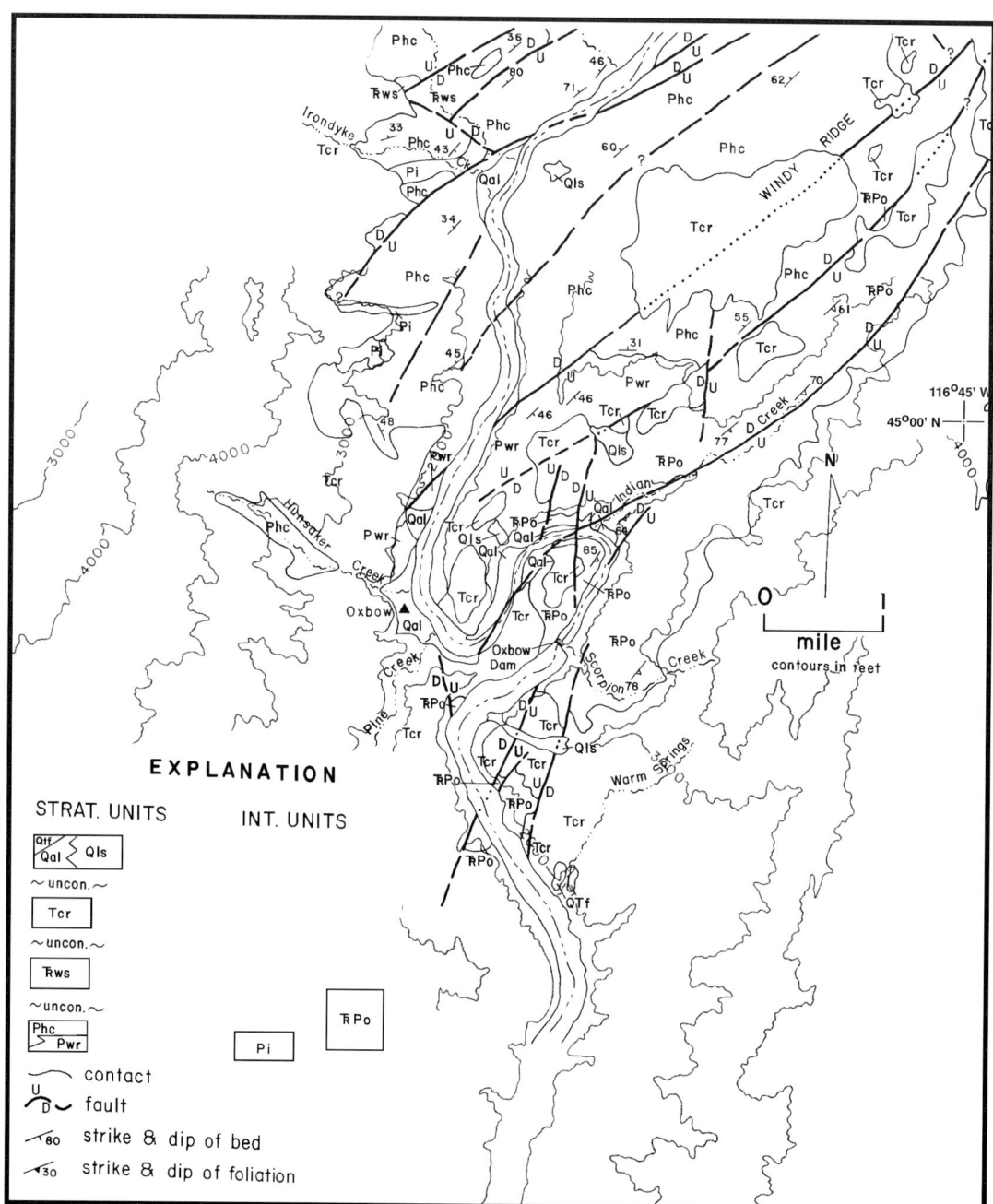

Figure 98. (page 106) Geologic map of the Oxbow area.
Geologic mapping completed by the author.

Figure 99. (page 107) Geologic map of the Homestead area.
Geologic mapping completed by the author.

Figure 100. (page 108) Geologic map of the Limepoint Peak to Kinney Creek area.
Geologic mapping completed by the author.

Figure 101. (page 109) Geologic map of the Kinney Creek to Hells Canyon Dam area.
Geologic mapping completed by the author.

Oxbow to Hells Canyon Dam 107

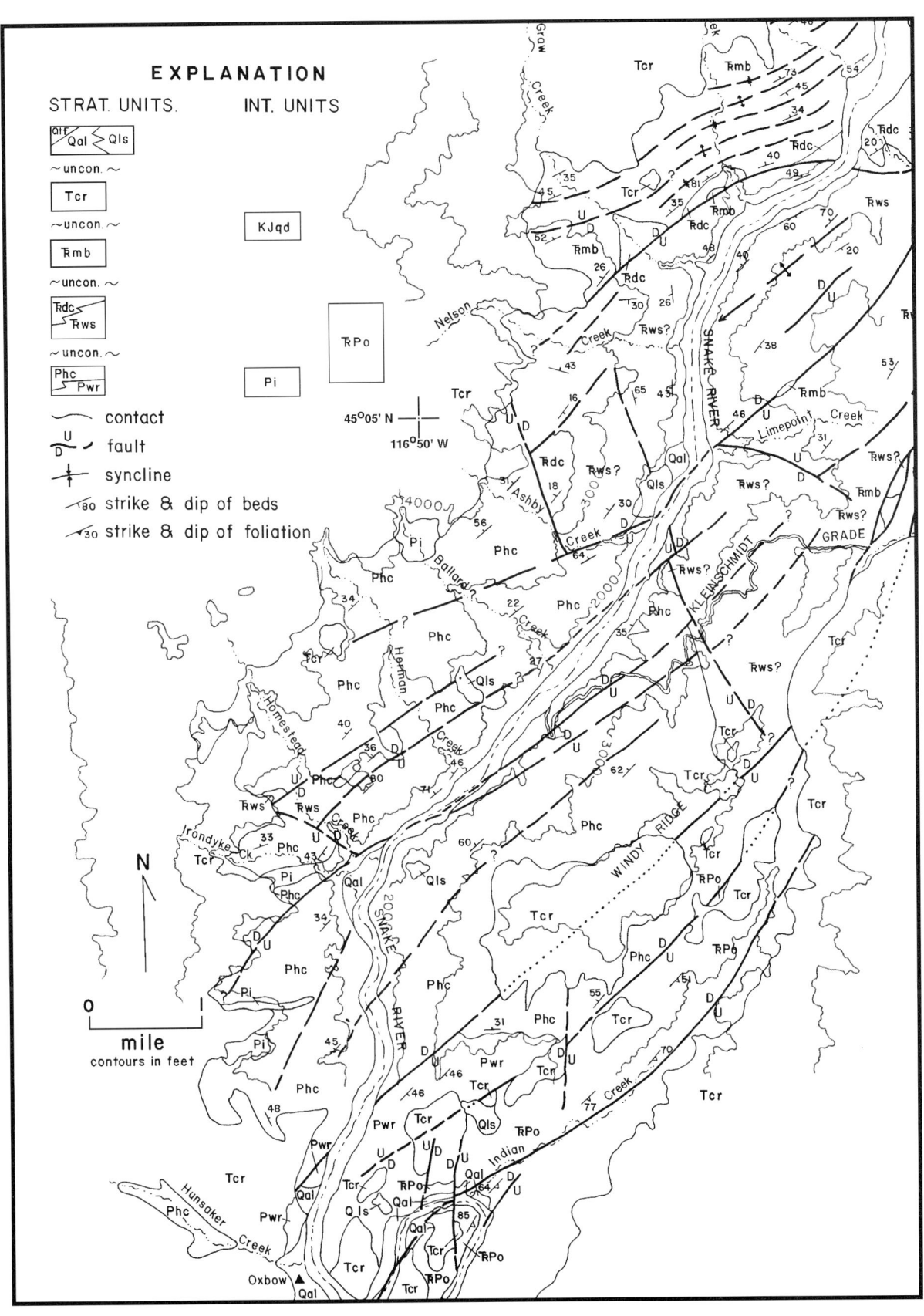

108 ISLANDS AND RAPIDS

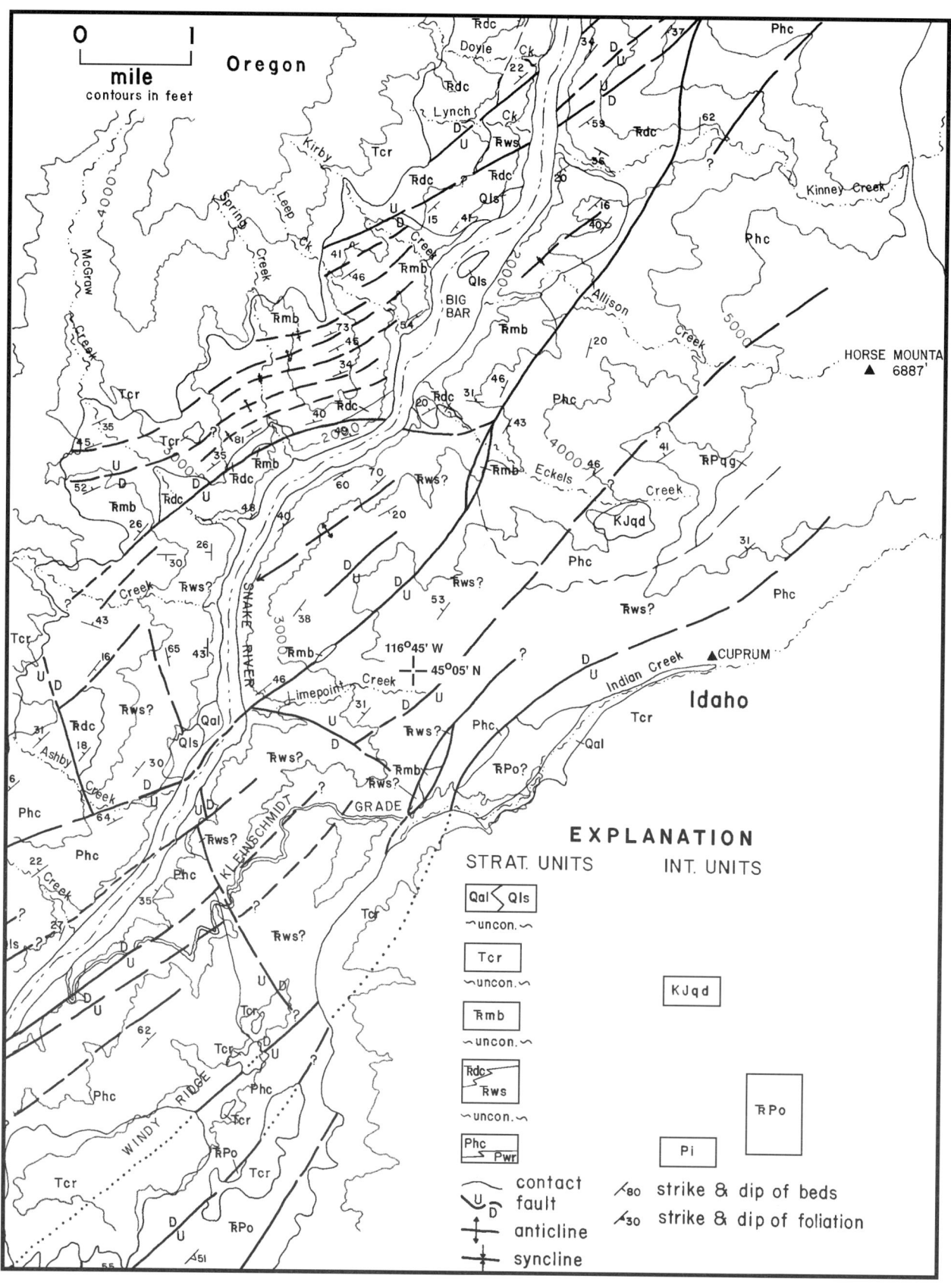

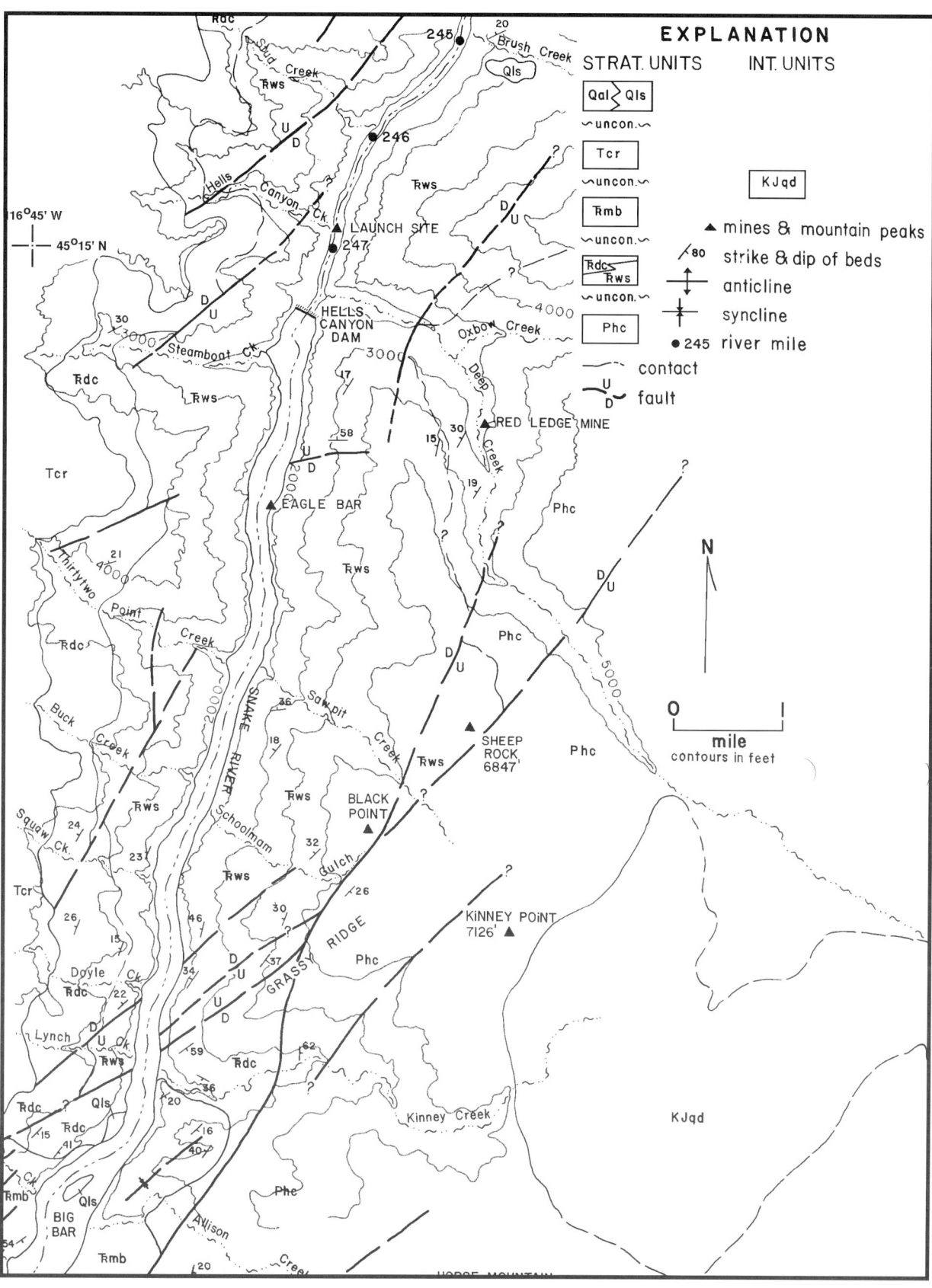

110 ISLANDS AND RAPIDS

Flows that crop out along the road near the surge tanks also have well-developed columnar jointing. Several flows contain large vesicles, some of which are filled with the green mineral **celadonite** and translucent and white zeolite minerals, including stilbite. Celadonite and zeolites also formed in the soil zone between lava flows.

Between the surge tanks and top of the incline, a basalt dike cuts the lava flows (Figure 104). Notice the textural differences between rocks in the flows and those in the dike. Thin glassy rims (selvage) mark the edges of the dike. These glassy rims were formed by the rapid chilling of hot basaltic magma when it was injected into the older—and cooler—rocks.

Continue along the road to the top of the hill. Stop there. Notice that as the road turns and continues to the south, it follows a low point in the rocks. This cut (or gap) most likely formed during a spillover from the Bonneville Flood about 14,500 years ago. Floodwaters that had been impounded behind the constricted channel of the Snake River finally overtopped the natural rock dam and cut a subsidiary channel. This low point is called a "wind gap" because wind presently blows through a gap that initially had been cut by water. Pre-Cenozoic rocks in the wind gap are part of the Oxbow Complex.

Oxbow Complex

The Oxbow Complex represents part of the Early Permian to Triassic root (or basement) to some of the Wallowa terrane strata. Most of the Oxbow Complex is composed of dikes. Many of these rocks—originally rhyolite, quartz diorite, trondhjemite, basalt, gabbro, and diabase—were sheared and subsequently recrystallized within a strong strain field to form mylonite, gneissic mylonite, and amphibolite. The original dikes probably crystallized at depths of 6 to 10 miles under the earth's surface. The rocks were raised to the surface by faulting, were deeply eroded, and then covered by lava flows of the Columbia River Basalt Group.

An excellent place to study these sheared and metamorphosed dikes of the Oxbow Complex is at the top of the hill in the wind gap where dikes dipping about 70° westward crop out on the north side of the road (Figure 105). The light-colored rocks are mylonites that were originally rhyolite dikes but subsequently were sheared and then recrystallized. Under the influence of a new strain regime, crystals of quartz and feldspar grew perpendicular to the principal stress direction and formed the foliations. The dark-colored amphibolites adjacent to the light-colored mylonites were originally basalt (and diabase) dikes. On close examination, one can see that these dark-colored rocks are also foliated. Based on the presence of metamorphic amphibole, the foliated dike rocks were heated to at least 500° C; this metamorphism, combined with the strain, changed the primary pyroxene minerals in the mafic rocks to amphibole and other secondary minerals such as epidote and chlorite. Metamorphism and strain also caused a strong foliation. These rhyolite and diabase dikes, probably the oldest rocks (Early Permian ?) in the Oxbow Complex, were subsequently intruded by Late Permian trondhjemite, quartz diorite, and gabbro and metamorphosed within a strong strain field in the Late Triassic.

Drive (or walk) down the jeep road that leads to the north from the wind gap, then walk east along the small road and trail that parallels the river. Many of the rocks retained their original igneous textures, but others were completely converted to mylonite, gneissic mylonite, and amphibolite. Some of the most spectacular rocks are the light green and blackish green gneissic

Figure 102. The Oxbow region. Notice how the river bends 180°. The origin of this bend is not well understood, but the pre-Cenozoic structure certainly affected the channel's strange configuration.

Figure 103. Looking south from a hill above Indian Creek. Notice how the river course on the left parallels the dikes that crop out at the apex of the bend.

Figure 104. Basalt dike and flows of the Columbia River Basalt Group that crop out along the road leading to the Oxbow Dam.

JULY, 1964. I STOOD on rocks of the Oxbow Complex for the first time. I was a second-year graduate student in geology at Oregon State University and had chosen part of the southeastern Wallowa Mountains near Halfway and Hells Canyon, from the Oxbow Dam to the site of the future Hells Canyon Dam, as part of a mapping assignment for a Ph.D. dissertation. I pounded on rocks in the wind gap and wrote in my notebook that I had no idea about either the rock types or what they might mean in a regional sense. I was thoroughly confused. And, all of the questions have not yet been answered.

Figure 105. Mylonite and amphibolite, well exposed in the wind gap west of the Oxbow Dam, were originally basalt (and/or diabase) and rhyolite dikes of probable Early Permian age. Outcrop width in the photograph is about 20 feet.

mylonites. The light green bands, or folia, are mylonitized quartz diorite; the dark bands are amphibole-rich and originally were mafic xenoliths in the quartz diorite.

From a vantage point near the wind gap, look north at a small keystone graben of Columbia River Basalt that was displaced downward along faults (Figure 106). Keystone grabens such as this generally indicate that the region has been under an extensional regime, such as that associated with pull-apart basins. Pine Valley near Halfway, Oregon, is an active pull-apart basin.

Return to the wind gap and proceed to the Oxbow Dam. Park on the west side of the dam where green and white trondhjemite dikes and greenish black mafic dikes crop out and extend northward past the dam along the spillway (Figure 107). The green color in the trondhjemite is caused by mineral grains of blue-green hornblende, light green chlorite, and pistachio-green epidote. The white and translucent minerals are feldspar and quartz, respectively. The trondhjemite that crops out near the dam has been radiometrically dated by the U/Pb method (Nicholas Walker) and crystallized about 249 Ma (latest Permian). Trondhjemite (and quartz diorite) dikes, such as those exposed here, cut older rocks elsewhere in the Oxbow Complex and are among the youngest dikes in the rock unit. Therefore, I presume that the 249 million-year-old age is the youngest for igneous rocks in the Oxbow Complex, and I suspect that rocks at least 260 million years old exist in the complex, particularly the mylonitic dikes noted above that have the same chemical composition as rhyolite dikes and flows within the Early Permian Hunsaker Creek Formation. An Ar/Ar radiometric age of metamorphic blue-green hornblende yielded a 214 million-year-old age. This Late Triassic age marks the time of latest metamorphism, and probably the age of shearing (during left-lateral movement) that formed most of the foliated rocks as well.

Walk or drive across the dam and then to its base on the east side. Cross the east spillway and observe the rocks exposed in the cliff face. The folded rocks, originally gabbro and diabase, are now amphibolite; they were

Figure 106. Graben with lava flows of the Columbia River Basalt Group that have been faulted down into pre-Cenozoic rocks. This outcrop is along the north side of the Snake River about one quarter of a mile west of the confluence of Indian Creek and the Snake River.

Figure 107. Trondhjemite dikes crop out along the spillway of Oxbow Dam. A sample from one of these dikes yielded a latest Permian U/Pb radiometric age.

deformed after shearing and metamorphism. Climb along the ridge north of Scorpion Creek where excellent exposures of the trondhjemite and amphibolite (formerly mafic igneous) dikes can be observed, along with their intrusive relationships. The trondhjemite dikes in Scorpion Creek are the younger.

Retrace the road to the intersection of Highway 86 and the old Baker-Homestead Highway. Turn right, drive past the campground and school and cross the bridge. Begin the following mile-by-mile log at the east end of the bridge.

The type stratigraphic section for the Permian Hunsaker Creek Formation is in Hunsaker Creek which lies west of the small village of Copperfield. It is approximately a mile walk up Hunsaker Creek to the first Permian outcrops. I measured and described about 2,400 feet of section along the north side of the creek. Volcanic sandstone and volcanic siltstone, many of which are tuffaceous, are the dominant rock types.

GUIDE TO THE GEOLOGY ALONG THE IDAHO POWER COMPANY ROAD

This part of the book uses odometer miles. Odometer mileage begins at the Snake River bridge and continues about 22 miles to the Hells Canyon Dam. The miles are given on the left and a text follows. Your car's odometer may not register exactly the same as the mileage given in this road log, but it should be close. I've driven the road with three vehicles, all of which gave somewhat different odometer readings. Within the text I've inserted the odometer reading at intersections of the road with creeks and other geographic features, which will help calibrate your vehicle's odometer readings with the road log. The geologic maps (Figures 98-101) also will help with orientation.

0.0 to 1.7 miles

Rocks directly above and east of the Snake River bridge are lava flows of the Columbia River Basalt Group (Imnaha Formation). A small outcrop of light beige Mazama Ash crops out near the road. After turning left onto the IPCO road, the first pre-Cenozoic rocks are tuffs and pyroclastic breccias of the Windy Ridge Formation. These blue-green rocks were explosively erupted from a volcano. Some of the rock fragments and crystals of feldspar and quartz were partly welded together by hot glass during the eruptions; subsequently, the glass changed to fine-grained feldspar and quartz during low-grade metamorphism. Angular-shaped rock fragments are clearly apparent on weathered surfaces. Abundant basalt, diabase, and gabbro dikes cut the bluish-green pyroclastic rocks.

No fossils have been recovered from the Windy Ridge Formation. The fragmental rocks in the Windy Ridge Formation are most likely related to some of the deposits that occur in the lower part of the Hunsaker Creek Formation. I suspect that the Windy Ridge Formation is also Early Permian.

At about 1.7 miles, a fault contact between rocks of the Windy Ridge Formation (on the south) and the Hunsaker Creek Formation runs from the road northeast across Windy Ridge. The depositional contact between the formations, as mapped along the east side of Windy Ridge, is gradational from coarse-grained tuffaceous strata of the Windy Ridge Formation to mostly epiclastic sandstone and volcanic breccia of the Hunsaker Creek Formation.

1.7 to 3.8 miles:

This section of the road traverses parts of the Hunsaker Creek Formation. The rocks, however, are not typical of most of the formation. For example, the first few outcrops are dark-colored volcanic breccia, a rock type that is uncommon in most other parts of the formation. Predominant rock types in the formation are light-colored pyroclastic breccia and tuff, with abundant epiclastic conglomerate, sandstone, and argillite.

This part of Hells Canyon is riddled with old mines and prospect pits. At mile 2.6, look across the reservoir at a small mine dump. Prospectors, particularly in the 1890s through the 1930s, scratched holes into the hillsides of the canyon looking for those elusive veins of gold. Across the river in Oregon at miles 3.4 to 3.5, south of Irondyke Creek, the bold outcrops on the hillside are thick quartz keratophyre (rhyolite) dikes.

At about 3.8 miles look southwest across the reservoir into Irondyke Creek (Figure 108). A few old buildings still remain from the town that was built mostly between 1910 and the middle 1930s. This small town of Homestead was spread across the alluvial fans of Irondyke and Homestead creeks. Many of the buildings in the town of Homestead were destroyed just before Hells Canyon Dam was completed because they would have been drowned by waters of the reservoir.

Several tailing piles or dumps related to the Irondyke Mine occur about half a mile west of the river along the walls of Irondyke Creek (Figure 109). Ore in the Irondyke Mine occurs as massive sulfide deposits that formed in hot springs on the Early Permian sea floor. The hot springs were associated with cooling magmas, some of which produced massive eruptions to form calderas and thick ash (tuff) deposits. In places, the springs deposited piles of debris rich in copper, gold, silver and other metals; the piles of metals and associated debris subsequently became oversteepened and failed, leading to landslides and debris flows that dispersed the deposit over the ancient sea floor. Modern examples of ores forming in this manner are found in the Okinawa Trough of Japan and in places along the oceans' spreading ridges.

3.8 to 5.3 miles

The road continues north through breccias and sandstones of the Hunsaker Creek Formation. Fossil brachiopods (*Megousia sp.*) of Early Permian age were found in rocks above this section of road and in the Irondyke Creek and Homestead Creek areas of Oregon. At about 5.0 miles, look across

---

THE VAUGHN brothers discovered the Irondyke Mine ore in 1896, but scant production was recorded until about 1910. A mill was located along the creek from about 1916 to 1928. From 1910 through 1934, approximately 35,000 ounces of gold, 256,000 ounces of silver, and 7,200 tons of copper were produced. The mine (present portal is slightly north of the buildings) was reopened in 1979 for 8 years. When it closed for the last time, the mine had produced another 20,000 ounces of gold, 40,000 ounces of silver, and 1,900 tons of copper. At the prices of $300 per ounce for gold, $5 per ounce for silver, and $1 per pound for copper, the ore extracted from the Irondyke Mine would be worth about $36 million.

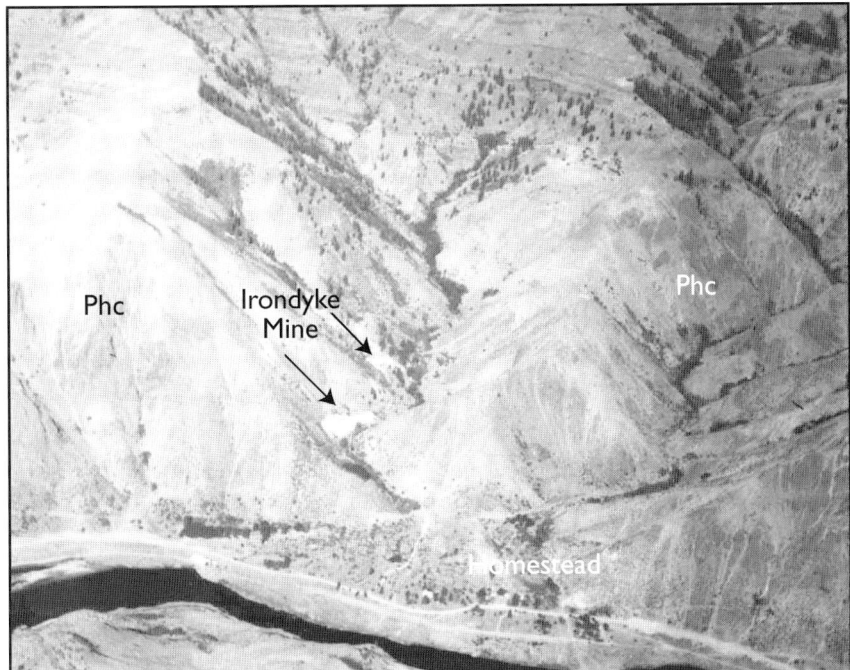

Figure 108. Irondyke Creek, Irondyke Mine, and old buildings that were built on an alluvial fan.

Figure 109. Irondyke Mine tailings near the main upper portal. A lower portal (not visible in this photograph) served as an entrance to the main tunnel during the latest major mining operation.

Figure 110. Breccia and sandstone beds of the Hunsaker Creek Formation near the mouth of Herman Creek.

the reservoir into the canyon of Herman Creek where breccia and sandstone occur in layered sequences along both sides of the creek (Figure 110). The debris in these beds are crudely to moderately graded in size from base to top, indicating that they were transported as debris flows and turbidity currents. Herman and Homestead creeks suffered severe blowouts during the winter of 1996-1997.

Figure 111. Tilted beds of the Hunsaker Creek Formation along the IPCO road near mile 5.3.

Figure 112. Rugged outcrops of the Hunsaker Creek Formation along Ballard Creek in Oregon.

5.3 to 5.4 miles:

A bold outcrop near mile 5.3 has beds dipping toward the reservoir at about 50° to 60°. The ancient sea floor is exposed on the tops of the tilted beds (Figure 111). Small one-inch-diameter holes in the rocks near the road were made when we drilled cores for paleomagnetic studies. These and other cores from the Hunsaker Creek Formation and the Wild Sheep Creek Formation indicate that the Permian rocks moved dramatically first south and then north over the earth's surface by plate tectonic processes before finally being accreted to ancestral North America.

5.4 to 6.4 miles

The road continues past Hells Canyon Park, which is maintained by the Idaho Power Company. At about mile 5.9, the Kleinschmidt Grade takes off toward Cuprum and Bear, Idaho. By following the road it is possible to drive to Council, Idaho, and Highway 95. The Kleinschmidt Grade, completed in 1891, was built by Albert Kleinschmidt to haul ore out of the Seven Devils Mountains to the river so that it could be hauled by steamboat upstream to Huntington, Oregon. The Kleinschmidt Grade is about 6 miles long, but the entire road from the foot of the grade through Cuprum and beyond can be followed for about 22 miles. The small town of Cuprum is approximately 10 miles from the IPCO road. Because the Kleinschmidt Grade is a single-lane road with turnouts, drivers should use extreme caution. However, the canyon views are spectacular and the drive is worth the extra effort. The foot of the Kleinschmidt Grade is about where the old Ballard Ferry was located until 1926. An interstate bridge constructed here linked Oregon and Idaho from 1926 until 1965. The bridge was destroyed during the construction of Hells Canyon Dam.

At mile 6.1, look directly across the reservoir into Ballard Creek (Figure 112). The bedded rocks on the north slope of Ballard Creek are tuff, conglomerate, breccia, sandstone, and siltstone of the Hunsaker Creek Formation with a few basalt and andesite dikes. I measured a stratigraphic reference section for the Hunsaker Creek Formation, about 1,800 feet thick, along the north side of Ballard Creek and paleomagnetic cores were collected by Bill Harbert and me from many of those outcrops.

Figure 113. Limepoint Peak rises high above the IPCO road. I suspect that the limestone (marble) was caught up along a strike-slip fault that is now inactive.

6.4 to 8.0 miles

This part of the road leads through volcanic breccia, sandstone, and finally (7.2 to 7.4 miles) bedded tuff of the Hunsaker Creek Formation. Triassic volcanic rocks overlie the Permian rocks along an unconformity that trends southeast beginning at about mile 8.0. Although the contact is sheared close to the road, farther uphill the unconformity can still be followed—but with great difficulty. Triassic rocks are mostly rusty red and black lava flows and volcanic breccia, whereas Permian rocks are light green and maroon tuff, breccia, and sandstone. From the road, look across the reservoir toward the Oregon side where a nearly vertical fault separates Permian rocks on the south from Triassic rocks on the north; this fault is well exposed along the road in Oregon near the mouth of Ashby Creek. The fault is marked by a zone of crushed rock measuring about 30 wide.

Continue driving along the road. Look straight ahead (north) at the large white rock outcrop that juts out prominently on the hillside above Limepoint Creek (Figure 113). This imposing outcrop, Limepoint Peak, is composed of marble. It is probably a metamorphosed fault sliver of the Martin Bridge Limestone. Several prospect pits and small mine portals dot the hillside below Limepoint Peak.

8.0 to 12.2 miles

The mouth of Limepoint Creek is about 8.7 miles north of the Snake River bridge. From here to mile 12.2 it is especially difficult to distinguish the Doyle Creek and Wild Sheep Creek formations. Fossils collected from the area between Limepoint and Eckels creeks indicate Middle and Late (Ladinian and Karnian) Triassic ages, although most fossils are Ladinian. It is apparent that the two formations are in part correlative; the Doyle Creek Formation in this area is composed of the oxidized and partly subaerial sections of the Wild Sheep Creek Formation and probably accumulated around volcanoes near and above sea level at the same time that flows of the Wild Sheep Creek Formation were erupted beneath the sea. The Wild Sheep Creek Formation is all submarine. In places, the Doyle Creek Formation overlies the Wild Sheep Creek Formation and in other places the two formations appear to interfinger. I designated these mixed rocks, those having both Wild Sheep Creek Formation and Doyle Creek Formation characteristics, as Wild Sheep Creek Formation on the map (Figures 99 and 100) because of the difficulty in separating them. The Doyle Creek Formation, however, can be separately mapped south of Nelson Creek where pumice-rich rust-colored tuff beds, breccias, and sandstones are abundant.

The remnant of a large slump and landslide occurs in Oregon, just south of Copper Creek. (There are 2 Copper Creeks, 2 Deep Creeks, 2 Cache Creeks, and 2 Big Bars in Hells Canyon.) The age of this landslide mass is not known. The slump and landslide feature can be traced southward nearly as far as Ashby Creek. The area between Ashby Creek and Copper Creek above the slump in Triassic rocks contains several abandoned mines, shafts, and prospect pits.

I MET THE TWO *Allwine brothers in 1964 at Copper Creek ranch. They had retired from the music business and showed me songs they had written for Metro Goldwyn Mayer in Hollywood. They were congenial and helpful, boasting a weedless garden full of goodies that they freely shared with visitors. At the time I was concerned with mapping the Permian-Triassic boundary and knew that both Permian and Triassic rocks occurred between the ranch and Ballard Creek. I explained my problem and the Allwine Brothers said they would solve it. I was asked to bring a Permian rock and a Triassic rock. Believe me, I was hooked, curious as a cat in front of a marble game. The next day I selected appropriate rocks and drove to the ranch. After a bowl of fresh cantaloupe, they asked me to bring out the rocks and to put my map next to the rocks. I placed the map, a green Permian tuff breccia, and a maroon Triassic basalt on their kitchen table. One of the brothers went outside and came back carrying a brass plumb bob that was tied to a fishing line. I watched with amusement as he wrapped the line around both hands. He held his hands about 8 inches apart so that the plumb bob dangled 2 feet below. He passed the plumb bob several times above the Permian rock and several times above the Triassic rock. Then he asked me to mark a pencil line on the map directly below the position of the plumb bob as it passed across the map. I made a line as the plumb bob moved slowly above the map. I forgot about the incident until sometime later, after I had finally figured out that the Permian and Triassic rocks in the immediate area were separated by the Ashby Creek fault. I looked at the line I had made following the plumb bob. The respective lines were nearly parallel and less than an inch apart.*

Figure 114. Folded Martin Bridge Limestone on the ridge north of Spring Creek (Color section, page 96).

Triassic epiclastic and pyroclastic debris hugs the IPCO road from about mile 8.7 to mile 12.2. Light gray outcrops of Martin Bridge Limestone cap the older volcanic rocks between McGraw and Spring creeks in Oregon and the beds are folded along Spring Creek (Figure 114). The thick Martin Bridge Limestone (about 1,500 feet measured along the south side of Kinney Creek in Idaho) is well exposed in Oregon and Idaho. At mile 10.2, look across the reservoir into McGraw Creek Canyon, and near mile 10.7 look into the canyon of Spring Creek. Both of these creeks were affected by blowouts in the Winter of 1996-1997. Water gushes from a hole in the limestone about one quarter of a mile upstream from the mouth of Spring Creek. Note that flows of the Columbia River Basalt Group overlie the Martin Bridge Limestone at high elevations in Oregon. The contact between these two formations is an angular unconformity—which means that the limestone beds were folded, faulted, and eroded before the basalt flows covered the older rocks. About 200 million years of earth history are represented by the hiatus between the formations. Hibble Gulch intersects the road at mile 12.0.

The contact between the Martin Bridge Limestone and the underlying coarse-grained volcanic breccia of the Doyle Creek Formation can be examined near the road at about mile 12.2. A major change in depositional settings is recorded. The older volcanic breccia of the Doyle Creek Formation was deposited in a dynamic setting. A thick debris-flow cascaded off the sides of an underwater volcano and was deposited in relatively deep water. Subsequently, the sea floor was raised and a shallow, more tranquil environment encouraged the deposition of calcareous plants and animals that settled out of the water column to accumulate as calcium carbonate (now limestone) beds of the Martin Bridge Limestone. No doubt, some time elapsed between the two depositional events, although I don't know how much.

12.2 to 14.8 miles

This 2.6 mile segment is exceptional because the rocks illustrate several significant aspects of the Hells Canyon geologic story. You are driving through the Martin Bridge Limestone. To the west the unconformity between the

Figure 115. Massive outcrops of the Martin Bridge Limestone (on right) along the south side of Kinney Creek.

Figure 116. General view of the Big Bar area during the construction of Hells Canyon Dam. Notice the gravel pit that was dug into Bonneville Flood deposits. The smoothed landslide deposit is shown by the letter "L." Curved rows of buildings are mobile homes that were occupied by the construction workers and their families. Jesse Smith's ranch buildings are in the trees west of the mobile home park.

limestone and flows of the Columbia River Basalt occurs above the high bench. Outcrops near Kinney Creek are imposing (Figure 115). In addition, the rounded top of a major landslide deposit rises above the reservoir at Big Bar (Figure 116). There is also some evidence of Bonneville Flood passage and an accumulation of Mazama Ash forms a thick deposit near the road. Water in the reservoir covers many of the features; fortunately, I made observations in the Big Bar area before the dam was completed.

Between miles 12.5 and 13.6, the road parallels the mostly submerged landform called Big Bar. Notice the rounded island protruding above water in the middle of the reservoir. This island, the top of a massive landslide deposit (Figure 117), was rounded off by the turbulent Bonneville Flood. The source area of the landslide can be observed on the canyon wall directly west of the island (Figure 118). The Big Bar deposit (Figure 119) is composed of mixed, and partly intertonguing, landslide debris, Bonneville Flood gravels and sands, and alluvial fan sediments.

Figure 117. The Big Bar landslide deposit before its submergence beneath the waters of Hells Canyon Dam reservoir *(Frontispiece Chapter 4, page 104)*.

Figure 118. Landslide scar in Oregon. The remaining rocks dip precipitously toward the river. Companion outcrops were dislodged, possibly during an earthquake, to form the landslide debris.

Figure 119. Deposits at Big Bar showing cobbles deposited by the Bonneville Flood and some overlying finer grained sands, probably deposited in slack water or a backeddy of that flood as the water backed up behind the narrow Hells Canyon stricture farther downstream.

IN MID-AUGUST *of 1964 I met Jesse Smith at his ranch on Big Bar. His ranch house now lies more than 100 feet below the reservoir's surface. Jesse was a character I admired because he personified so much of what life in the canyon was all about. Jesse loved Big Bar. He had homesteaded there around 1910, and the future slack water (behind Hells Canyon Dam) was forcing him to leave. Every time I see the island in the middle of the reservoir at Big Bar, I think about Jesse and other homesteaders who spent most of their lives scratching out a living on land they loved.*

I stopped at the ranch house and asked permission to walk across his land. He served me coffee from grounds so thick in the pot that I'm sure he just dumped in fresh Folgers each day without cleaning out the old grounds. He also served hard biscuits and grape jelly. After coffee I asked if I could walk across his land for a few days. He assured me that it was okay.

The next morning I drove past his house and honked the horn, to let him know I was in the area, drove to the north end of Big Bar, climbed over a fence, and started across the field. After walking less than 100 yards I saw (and heard) a pickup rumbling down the road. It stopped behind my car. Jesse jumped out with a rifle in his hands. He put the rifle across the hood of his pickup and pointed it in my direction. "What are you doing here?" he yelled. I dropped the hammer, maps, and notebook, raised my hands, and walked toward him. When I got close, he asked again why I was there, and I reminded him of our conversation the previous day. His eyes lit up in recognition as he lowered the rifle. Then he walked around the pickup toward me. "I thought you were one of those power company guys," he said. "They've been walking all over my land driving stakes and taking things." He then added, "They're stealing my land, you know."

An archeological excavation was completed at Big Bar in 1963. The site was just north of the Snake River—Allison Creek confluence about 30 feet above river level. Evidence for "Early" and "Late" occupancy suggests that the site had been occupied for several hundred years.

Between miles 12.8 and 13.2, look across the reservoir and notice the flat-lying flows of the Columbia River Basalt Group overlying folded Martin Bridge Limestone above the angular unconformity. Large well-rounded boulders, eroded from more ancient bedded quartzite units in central Idaho and western Montana, lie along the unconformity (Figure 120). The boulders, some of which measure more than 20 inches in diameter, were transported by a powerful river that is unrelated to the present-day Snake River. The

Figure 120. Large quartzite boulder of probable Precambrian (or early Paleozoic) age that is part of a boulder field that rests on top of the surface of the Martin Bridge Limestone between McGraw and Spring creeks.

Figure 121. Folds in the Martin Bridge Limestone above Leep Creek in Oregon *(Color section, page 96)*.

Figure 122. Mazama Ash along the IPCO road near Allison Creek.

presence of these boulders tells geologists that the mountains to the east were very rugged and high at some time between the erosion of the Martin Bridge Limestone surface and the eruption of the lava flows that compose the Columbia River Basalt Group.

At about 13.2 miles stop near Allison Creek. Spectacular folds within the Martin Bridge Limestone can be seen to the southwest in Oregon (Figure 121). Some axes of folds in the Martin Bridge Limestone merge into faults within the underlying Doyle Creek Formation. The structural style from fold to fault changes with the competency of the beds.

A thick pile of Mazama Ash (Figure 122) butts up against the west side of the road near mile 13.2, just south of where Allison Creek flows through a culvert under the road. This is one of the thickest deposits of Mazama Ash in the canyon. The ash was deposited during and immediately after the eruption of Mt. Mazama (the mountain that now holds Crater Lake in the Cascade Mountains) approximately 6,850 years ago. A thin layer of volcanic ash, similar perhaps to a one-inch snowfall, clothed the surrounding terrain during the eruption. The next major rainfall washed the ash into the streams, and subsequently deposited thick layers of ash on alluvial fans where stream velocities suddenly decreased and ash-clogged waters overtopped the pre-existing channels. Special conditions, such as a shift in the stream course and the rapid formation of another major channel, allowed the preservation of some volcanic ash layers on the alluvial fans, like this one near the mouth of Allison Creek.

Large boulders, deposited by the Bonneville Flood, were recovered from borrow pits about half a mile southwest of this site (Figure 119), crushed to gravel size, and then used in the concrete of Hells Canyon Dam. The stair-step terraces that can be seen just north of this stop, between the road and reservoir water level, were the sites for mobile homes that had been brought into the canyon for workers and their families during the construction of Hells Canyon Dam.

Figure 123. Massive cliff of Martin Bridge Limestone between Allison and Kinney creeks. The limestone here is about 1,500 feet thick. Kinney Creek Rapids are in the foreground.

Figure 124. Large clasts of volcanic rocks make up a Doyle Creek Formation breccia that occurs beneath the Martin Bridge Limestone at the Kinney Creek pull off.

About one half of a mile from the road along the north side of Allison Creek, and about 100 feet above it, is the Allison Creek shelter. It was excavated by archeologists in the late 1960s. This rock shelter is a cave in the limestone. The cave has been explored to a depth of 200 or 300 feet. Caves and **sinkholes** are rare in the limestone, but some small sinkholes do occur on the high platform between Allison and Kinney creeks. A rock that I dropped into a sinkhole on that platform fell for more than 2 seconds before hitting bottom.

Cliffs of Martin Bridge Limestone tower above Kinney Creek (Figure 123). The limestone body exposed in the Big Bar area was faulted (and folded) downward into the older rocks and thereby was preserved while equivalent limestone was eroded from the surrounding terrain. The limestone is about 1,500 feet thick here and formed on a shallow platform in warm waters. No reefs have been reported from limestone in the Big Bar area, but one has been described in the southern Wallowa Mountains near Summit Point southwest of Cornucopia. Four new species of fossils were described by Dr. Cathy Newton of Syracuse University from a diverse suite of silicified invertebrate fossils recovered from outcrops of the Martin Bridge Limestone on the ridge north of Spring Creek.

Stop at the pullout just south of Kinney Creek (mile 14.8). A volcanic breccia makes up the top beds of the Doyle Creek Formation (Figure 124). The sheared contact between red and green volcanic breccia of the Doyle Creek Formation and the Martin Bridge Limestone can be observed along the road about 200 feet south of this stop. Along Kinney Creek, just above the pull off, the transition between the two formations consists of about 3 feet of pale green calcareous sandstone and siltstone. There is no non-calcareous (terrigenous or volcanic) detritus in the insoluble residues of more than two dozen rocks from the Kinney Creek section of Martin Bridge Limestone that I studied. These data indicate that this part of the Martin Bridge Limestone was probably deposited far from volcanic islands.

IN 1965 BEFORE *Hells Canyon Dam was completed, I crossed the river above and below the Squaw Creek and Buck Creek rapids in a small four-man raft. I carried the raft from the IPCO road to river level, where I inflated it, and then rowed across the river. One day in mid-August I rowed across the river above Squaw Creek Rapids, climbed the chute just below the rapids, and spent most of the day mapping in the upper levels of the canyon. I started down the walls around 7:00 p. m., but couldn't find a safe way. It is much easier to climb up the steep slopes of a canyon than it is to descend. Every time I started down, a steep cliff thwarted my progress. Finally, dusk descended and I stayed the night sitting on the rocks. It was a long night. The next morning I trekked back to the chute, where I had climbed up and finally arrived at camp before noon. My family was anxiously waiting.*

Figure 125. Spheroidally weathered dike of Columbia River Basalt exposed along the IPCO road at about mile 16.3.

Figure 126. Black Point area before construction of the Hells Canyon Dam.

14.8 to 16.5 miles

This section along the IPCO road traverses the area from Kinney Creek to the Black Point pullout. The major rock types are volcanic breccia and sandstone of the Wild Sheep Creek Formation. Pistachio green outcrops along the road at mile 15.2 are epidote-rich breccia. This abundance of epidote is associated with a wide shear zone. The red coloration on some of these outcrops is caused by the mineral hematite.

Between miles 15.8 and 16.2, sandstone beds of the Wild Sheep Creek Formation are especially well exposed. Specimens of *Daonella beedi*, a Middle Triassic flat clam, were recovered from outcrops above the road. Paleomagnetic samples were selected from some of these outcrops. At about 16.3 miles a dike of Columbia River Basalt crops out along the east side of the road. The basalt has a spheroidal appearance caused by differential weathering along joints (Figure 125).

At mile 16.5, stop at the Black Point pullout (Figure 126). A black volcanic breccia, well exposed along the road near the pull out, is representative of many rocks in the Wild Sheep Creek Formation in this part of Hells Canyon. The creek to the southwest in Oregon is Squaw Creek. Squaw Creek Rapids, now covered by water, was one of the largest in the canyon. The river level dropped 17 feet over a distance of approximately 700 feet.

16.5 to 20.0 miles

Rocks along the road between Black Point and the abandoned IPCO site at Eagle Bar consist almost entirely of the Wild Sheep Creek Formation. (The administrative buildings were sited here during construction of the Hells Canyon Dam.) A quartz diorite dike cuts the Wild Sheep Creek Formation around the first turn north of Black Point. Dikes of this composition commonly cut volcanic rocks of the Seven Devils Group. They are probably either Late Triassic or Early Jurassic in age. At about mile 17.3, Buck Creek Canyon can be seen in Oregon. Buck Creek Rapids was often described by boatmen as the most awesome set of rapids in the canyon. At about mile 18.4, Thirty-two Point Creek enters the reservoir in Oregon. The rocks on the Idaho side dip more steeply than those on the Oregon side, possibly because the river follows a fault line.

Towering cliffs hug the road between miles 19 and 20. These rocks are mostly thick beds of volcanic breccia and they dip toward the road and reservoir. At about mile 20, a trail takes off up the hill in Idaho. This trail leads to Red Ledge Mine in Deep Creek, a distance of about 6 miles from the IPCO road.

The trail to the Red Ledge Mine has not been maintained and is washed out in places. Although a round trip can be made in one day, I suggest an overnight stay near the mine before returning. The Red Ledge Mine is sited in what is most likely part of a tilted Permian caldera. The ore minerals are incorporated within a massive sulfide deposit that had a submarine springs origin similar to that of the Irondyke Mine near Homestead. The Red Ledge Mine consists of twenty-three patented claims and a large number of unpatented claims covering about 1,500 acres. No ore has been sold although at least a million dollars was spent on exploration, tunneling, road building, and drilling. I doubt if the mine will ever be exploited because of its isolation and the fact that it is surrounded by a Wilderness Area. If you hike to the mine, keep your eyes open for bears and rattlesnakes.

20.0 to 21.6 miles

The Wild Sheep Creek Formation crops out along the road between Eagle Bar and Hells Canyon Dam. Near mile 21.2, the mouth of Steamboat Creek can be seen in Oregon. It is the last major creek before Hells Canyon Dam. The creek's name is derived from the *Shoshone*, a paddle wheeler that was tied up near the creek's mouth during the winter of 1869-1870.

At about mile 21.4, lava flows and sills with **glomeroporphyritic** textures crop out in the cliffs and in boulders along the road (Figure 127). These peculiar rocks are common from here north along the road to the dam and from thence down the trail to Deep Creek. The large white to light green **phenocrysts** (**glomerols**) are plagioclase feldspar. Some form simple crosses, but others are radiating accumulations of crystals ranging from one to two inches in diameter. These large crystals crystallized first in a mostly

OCTOBER, 1985. *Trudy and I were camped in Deep Creek and working out of the Red Ledge Mine building site, where we had camped for about a week. One of those days, October 5 to be exact, we were mapping the Permian-Triassic unconformity along the west side of the Deep Creek canyon. Trudy was nervous because we had heard a bear outside our tent the night before and had seen a large black bear on the ridge east of camp that morning. She was ahead of me about 30 feet and suddenly stopped. "Rattlesnake," she shouted. "There's another one." I hadn't even reached her side when she screamed, "And there's one over there." We had stumbled onto a nest. Rattlesnakes were congregating to "ball up" for the winter. We stood together and counted 16 snakes before retreating. Most were rattling ominously. Trudy was shaken as we carefully walked among the rocks and trees the rest of that day. She spent a lot of time scanning distant ridges and arroyos for bears and scouting nearby rocks and crevices for snakes.*

Figure 127. Some flows and sills of the Wild Sheep Creek Formation have glomeroporphyritic textures. These outcrops are along the IPCO road at about mile 21.4.

HELLS CANYON *Dam and other dams upstream have affected the fragile environments in Hells Canyon. The total effect of dams on the canyon, both beneficial and detrimental, is impossible to measure. The dams are important to the region's economy; yet they have caused irreversible damage to the beaches, which no longer caress the sides of the river between Hells Canyon Dam and the Salmon River. The dams have also impeded the migration of anadromous fish. Without the intervention of the Federal Government in the late 1960s and early 1970s, the remaining 60 miles of free-flowing river would be flooded by a succession of at least 3 additional dams. A river and its free spirit, once gone, can never be returned.*

liquid magma chamber several miles underground and were carried to (flows) and near (sills and dikes) the surface during volcanic eruptions.

Hells Canyon Dam Area

Hells Canyon Dam is about 21.6 miles north of the bridge that crosses the river near Oxbow. It marks the end of this road guide. I was working in the canyon during the construction of Hells Canyon Dam (Figure 128) and visited both the 1800-foot-long diversion tunnel and the stripped off area where the 320-foot-high dam was built. Cement was poured beginning in March, 1966, and the reservoir was completely filled in 1968. The dam's spillway is designed to handle 300,000 cfs of water flow.

About 100 feet south of the dam along the IPCO road, Columbia River Basalt dikes cut the Wild Sheep Creek Formation. One of the dikes bifurcates toward the top of the outcrop (Figure 129). On the east side of the dam a trail takes off to Deep Creek. The walk is worthwhile because there are many different porphyritic textures in the Triassic lavas. Furthermore, the trail passes the diversion tunnel and then goes on to Deep Creek where tired feet can be dangled in the water.

Drive across the dam and follow the road to the visitor's center above the boat launch (mile 22.8). An archeological excavation site can be seen at the bottom of the stairs. Take a short hike along the riverbank for about one half of a mile and you will have some idea of what the canyon was like before the dam was built. The rocks at the visitor's center and along the trail are dark green and maroon lava flows and volcanic breccias. All are part of the Wild Sheep Creek Formation.

While at the visitor's center, look across the river and about 800 feet up the canyon wall to see a cathedral-like outcrop that hangs menacingly over the river. A large part of the outcrop has already slipped down into the canyon as evidenced by the talus cone. This outcrop is a geologic hazard, but only one of several that occur in the canyon. The strata are dipping about 30° toward the river and, in the spring, water percolates beneath the towering mass onto the exposed bedding plane. The rock mass could easily slide into the canyon, particularly if slope failure is encouraged by shaking during an earthquake.

Oxbow to Hells Canyon Dam 127

Figure 128. Hells Canyon Dam during construction in 1965. A small dam was used to send water through a diversion tunnel that empties on the right. White water marks the outlet of water from the tunnel.

Figure 129. Columbia River Basalt dike cutting the Wild Sheep Creek Formation just south of Hells Canyon Dam.

Annotated Bibliography

I cannot list all of the references here that are necessary for an adequate understanding of the Hells Canyon region and the tectonics and other processes that formed it. Rather, I've listed those that include enough specific and general details to satisfy most questions that might be asked by inquiring students. Below each of the references I've written a few comments to help the reader focus on those books and articles that may be of further interest. The reference style is modified somewhat from that used by the American Psychological Association.

Aikens, C. M. (1986). Archeology of Oregon. Washington, D.C.: U. S. Department of Interior, Bureau of Land Management, U. S. Government Printing Office.
> Aikens presents an excellent review of the different Indian cultures in Oregon. Although the author does not discuss Hells Canyon archeology, he does introduce the reader to human migration into North America (25,000—15,000 years ago during maximum continental glaciation when a wide Bering land bridge existed). Some archeologists believe that the Paleo-Indians entered the Oregon country as early as 15,000 years ago, but certainly by 11,000 years ago. Archeologists also speculate that the Paleo-Indians had hunted the mammoth, horse, camel, giant bison, and several other species to extinction by about 9,000 years ago.

Alt, David, & Hyndman, D. W. (1995). Northwest Exposures: A Geologic Story of the Northwest. Missoula, Montana: Mountain Press Publishing Company.
> The authors describe the geology of the Pacific Northwest in an easy to read format. Many sketches and illustrations complement the text. I recommend this book for a regional understanding of the Pacific Northwest.

Ashworth, W. (1977). Hells Canyon: the Deepest Gorge on Earth. New York: Hawthorne Books.
> Ashworth reviews several aspects of Hells Canyon, including its name (p. 65-66). The names Box Canyon, Seven Devils Gorge, and Grand Canyon of the Snake all were used at various times before the name Hells Canyon was finally accepted by the Oregon Geographic Names Committee in December of 1970.

Bacon, C. R. (1983). Eruptive History of Mount Mazama and Crater Lake Caldera, Cascade Range, U. S. A. Journal of Volcanology and Geothermal Research, 18, 57-115.
> Bacon focuses on the eruptive history of Mt. Mazama and chronicles the climactic eruption at 6,845 +/- 50 years before present that created the caldera now holding Crater Lake. I've used 6,850 years before present in this book, but Mt. Mazama may have erupted climactically as early as 6,895 years ago or as late as 6,795 years ago. Radiocarbon dating of charcoal that was found within the Mazama Ash was the method used to obtain the date of eruption. The +/- indicates the uncertainty of the age based on the radiocarbon dating methods.

Bingham, R. T., & Henderson, D. M. (1980). Guide to the Common Plants of Hells Canyon. Washington, D. C.: U. S. Department of Agriculture, Forest Service, U. S. Government Printing Office.
> This excellent guide to 100 plants in the Hells Canyon region can be tucked in your back pocket. It is well illustrated. I recommend it to anyone who spends more than a day in the canyon. The black and white illustrations can be colored to match the color of the various wild flowers. Unfortunately, at the time of this writing the book is out of print.

Bishop, E. M., & Allen, J. E. (1996). Hiking Oregon's Geology. Seattle, WA: The Mountaineers.
> What a joy it is to read this book and to follow some of the trails that the authors recommend. Some trails in Hells Canyon are suggested in Chapter 10. The book is well laid out and illustrated, easy to read, and written by scientists who thoroughly understand the geology of Oregon.

Bonnichsen, W., & Breckenridge, R. M. (Eds.). (1982). Cenozoic Geology of Idaho. Moscow, ID: Bureau of Mines and Geology, Bulletin 26.
> This volume contains forty-two separate articles on various aspects of Idaho's Cenozoic geology. The articles were written for professional geologists, but interested amateur geologists can also benefit from reading them. Of particular interest are the articles on the Columbia River Basalt Group (Chapter two) and on the Pleistocene Flood Deposits and Gravels (Chapter thirteen).

Boreson, K. E. (1976). A Bibliography of Petroglyphs/Pictographs in Idaho, Oregon, and Washington. Northwest Anthropological Research Notes, 10, 123-146.

> Boreson provides an excellent review of petroglyphs and pictographs in the Pacific Northwest. Maps show the approximate locations of sites in Hells Canyon. Boreson is a recognized expert on Hells Canyon prehistoric art. The archeologist for the Hells Canyon National Recreation Area, stationed in Enterprise, Oregon, can refer the reader to other articles of interest.

Brucker, R. H. S., & Evans, Brock. (1970). In the Matter of the Joint Applications for License for Middle Snake River Project in the Middle Snake River in the States of Oregon and Idaho by Pacific Northwest Power Co. and Washington Public Power Supply System: Projects No. 2243 and 2273. Seattle, WA: Brief of Sierra Club, Federation of Western Outdoor Clubs, and the Idaho Alpine Club.

> Many of the issues raised by these authors have not been adequately resolved. I recommend this brief to anyone interested in the historical aspects of dam construction and the efforts that were focused on keeping additional dams out of Hells Canyon.

Caldwell, W. W., & Mallory, O. L. (1967). Hells Canyon Archeology. Washington, D. C.: Smithsonian Institution, Publications in Salvage Archeology, 6.

> The authors review archeological investigations of sites at Robinette, Connor Creek, Big Bar, the Allison Creek Cave, Somers Creek, Dry Creek, Divide Creek, and Big Canyon.

Carrey, Johnny, Conley, Cort, & Barton, Ace. (1979). Snake River of Hells Canyon. Cambridge, ID: Backeddy Books.

> This book is a "must buy" for visitors to Hells Canyon. It is full of history and humor. The authors have a deep respect for Hells Canyon. Ace Barton grew up in the canyon and a new plant species, the Bartonberry (Rubus bartonianus), was named in honor of his mother. The data for this book were well researched; some of the historical incidents and statistics printed in their book are used in several sidebars of my text.

Clegg, Edith. (1984). Rattlesnakes and Rapids: A Woman's Journey Against the Current in 1939. Idaho Yesterdays, 28, 10-19.

> This is an amazing account of a woman and four boatmen, who, with wooden boats, traversed the Snake River Canyon from its mouth to Huntington, Oregon. The journal notes (May 11 through June 5, 1939) about the trip between Lewiston, Idaho, and Huntington, Oregon, are particularly interesting for Hells Canyon enthusiasts.

Committee on Interior and Insular Affairs (Henry Jackson, Chairman). (1970). Hearing Before the Subcommittee on Water and Power Resources of the Committee on Interior and Insular Affairs, United States Senate, Ninety-First Congress, Second Session on S. 940, a Bill to Prohibit the Licensing of Hydroelectric Projects on the Middle Snake River Below Hells Canyon Dam for a Period of Ten Years. Washington, D. C.: U. S. Government Printing Office.

> These proceedings (February 16, 1970) contain statements, testimonies, and letters on the subject of a moratorium on building dams in the middle Snake River. It is worth reading because of the many issues, some still unresolved, that were discussed when building the dams on the Snake River. Within Floyd Harvey's testimony, the letter of Arthur Godfrey (p. 124-126) was submitted that had been written to the Secretary of Interior, Walter Hickel, concerning the need to keep additional dams out of Hells Canyon. Passions were high in many of these testimonies and letters.

Earl, Elmer. (1990). Hells Canyon A River Trip. Lewiston ID: Lewiston Printing.

> Captain Earl has some fascinating stories about the Snake River and Hells Canyon told in a homespun, no nonsense manner. The book is well worth the investment. He provides the reader with unique historical sketches. His book, however, should not be used as a geologic guide. In fact, I hereby promise Elmer that I won't write about navigating the Snake River by power boat if he won't write about the geology of Hells Canyon.

Evans, Brock. (1968, September). Hells Canyon on the Snake. Sierra Club Bulletin, 53, 7-11.

> The author wrote a motivational article on the controversy over building dams and the need to stop further construction. It was written after Hells Canyon Dam was constructed and before Congress established the Hells Canyon National Recreation Area.

Gilbert, G. K. (1890). Lake Bonneville. Washington, D.C.: U.S. Geological Survey, Monograph No. 1.

> Gilbert's book draws attention to the great size and geologic history of Lake Bonneville. The book also is complementary to the books by Harold Malde (cited below) and James O'Connor (cited below) and was the first building block in our geologic understanding of Lake Bonneville and the Bonneville Flood.

Goldstrand, P. M. (1987). The Mesozoic Stratigraphy, Depositional Environments, and Tectonic Evolution of the Northern Portion of the Wallowa Terrane, Northeastern Oregon and Western Idaho. Bellingham, WA: Western Washington State University, Masters Thesis.
Goldstrand discusses the region between Mountain Sheep Creek and the Oregon-Washington boundary. He does an excellent job of separating units within the Wild Sheep Creek Formation and also of distinguishing mappable units within the Coon Hollow Formation. For anyone interested in the more technical aspects of the geology in that region, this thesis is an excellent reference. I incorporated some of his interpretations in the maps of the lower parts of Hells Canyon below the Imnaha River.

Hooper, P. R., & Swanson, D. A. (1990). The Columbia River Basalt Group and Associated Volcanic Rocks of the Blue Mountains Province. In Geology of the Blue Mountains Region of Oregon, Idaho, and Washington: Cenozoic Geology of the Blue Mountains Region (pp. 63-99). Washington, D. C.: U. S. Geological Survey, Professional Paper 1437, U. S. Government Printing Office.
The authors succinctly describe and interpret rocks of the Columbia River Basalt Group based on their many years of experience with the lava flows and the regional geology. Some of their interpretations have been incorporated in this book.

Irving, Washington. (1961). The Adventures of Captain Bonneville, U. S. A., in the Rocky Mountains and the Far West. Norman, OK: University of Oklahoma Press.
The story of Captain Bonneville's attempt at traversing the Hells Canyon country is chronicled here. He was an adventurous Cavalry officer. I am relieved that I did not have to serve under his command.

Jordan, Grace. (1954). Home Below Hells Canyon. New York: Thomas Crowell Company (Reprinted by University of Nebraska Press, Lincoln).
Grace Jordan lived at Kirkwood Bar, about five miles above Pittsburg Landing, from 1933 to 1944. This is a story about the Jordan family and their neighbors in the canyon and surrounding areas. Len Jordan later became governor of Idaho and afterwards a U. S. senator. The vibrancy of people, both during the depression and the two or three decades afterwards, is far different from the present. Where once residents could barely scratch out a living to survive, people now focus on rafting, powerboating, fishing, and other recreational activites.

Josephy, Alvin M., Jr., and others. (1983). The Nez Perce Country. Washington, D. C.: U. S. Superintendent of Documents, National Park Service Handbook 121.
Josephy's book provides a very short history of the Nez Perce people and the events that affected them. It also is a guide to the Nez Perce National Historical Park.

Leen, Daniel. (1988). An Inventory of Hells Canyon Rock Art. Enterprise, OR: Final Report to Hells Canyon National Recreation Area, Wallowa- Whitman National Forest, U. S. Forest Service.
This report is available from the Wallowa-Whitman National Forest, Baker City, Oregon. It contains drawings and maps of all sites inventoried. Remember not to disturb, in any way, the rock art in Hells Canyon. Look, take pictures, but don't touch. Rock art cannot be replaced.

Leopold, Luna B. (October, 1969). Landscape Esthetics. Conifer, 10, 12-21.
Leopold compared the esthetic characteristics of the Snake River and Hells Canyon with the esthetic characteristics of eleven other rivers and their canyons in an objective manner using forty-six factors to arrive at five "evaluation numbers." From these "evaluation numbers," he derived the "scale of valley character" (height of nearby hills + width of valley=landscape scale; landscape scale + scenic outlook=landscape interest; landscape interest + degree of urbanization="scale of valley character"). Hells Canyon ranked first in esthetics and uniqueness. He then compared the Snake River at Hells Canyon with the Merced River in Yosemite, Snake River in Grand Teton National Park, the Yellowstone River in Yellowstone National Park, and the Colorado River in the Grand Canyon by plotting "scale of valley character" versus "scale of river character." Hells Canyon and the Snake River, by the most objective methods of evaluation, rank very close to the Grand Canyon and Colorado River in esthetics. Hells Canyon is indeed a national treasure and resource.

Malde, H. E. (1968). The Catastrophic Late Pleistocene Bonneville Flood in the Snake River Plain, Idaho. Washington, D. C.: U. S. Geological Survey Professional Paper 596, U. S. Government Printing Office.
Malde's book is a classic on the Bonneville Flood and shows the erosional effects along parts of the Snake River in the Snake River Plain region below American Falls. He does not discuss the flood's effects in Hells Canyon.

McArthur, L. W. (1982). <u>Oregon Geographic Names</u>. Portland, OR: Oregon Historical Society.

On page 357 McArthur noted that the name Hells Canyon was applied first to the river approximately between the site of the present Oxbow Dam to the mouth of the Grande Ronde River, which is the same definition I applied in this book. The Geographic Names Committee recognized the name Hells Canyon in December, 1970, when they voted to change the creek's name of "Hells Canyon" (near the visitor's site below Hells Canyon Dam) to "Hells Canyon Creek." Apparently, this volume was compiled after Mr. McArthur's death because the compiler wrote that McArthur had fought the name change.

Morrison, R. (1963). <u>Pre-Tertiary Geology of the Snake River Canyon Between Cache Creek and Dug Bar, Oregon-Idaho Boundary</u>. Eugene, OR: University of Oregon, Ph. D. Dissertation.

This Ph.D. dissertation is the first written about the rocks in part of Hells Canyon and a classic for field geology in the region. Dr. Morrison worked the river with a small aluminum boat and a 25 hp motor. Imagine threading your way through the Imnaha Rapids with such a small boat after a difficult day climbing the canyon walls.

Norton, Boyd. (January, 1970). The Last Great Dam. <u>Audubon, 72</u>, 14-27.

This article is mainly a history of dams and what the power companies were trying to accomplish in the late 1960s.

O'Connor, James. (1993). Hydrology, Hydraulics, and Geomorphology of the Bonneville Flood. Boulder, CO: <u>Geological Society of America</u>, Special Paper 274.

O'Connor quantifies the effects of the Bonneville Flood and the book contains maps of the deposits in Hells Canyon. Anyone interested in flood dynamics will want to read this book.

Orr, E. L. & Orr, W. N. (1996). <u>Geology of the Pacific Northwest</u>. San Francisco, CA: The McGraw-Hill Companies, Inc.

Everyone interested in a comprehensive review of the geology of the Pacific Northwest should have a copy of this outstanding book. It is written for students of geology, both amateurs and professionals. The figures and diagrams are well done. I certainly recommend this book, both for easy reading and for your bookcase.

Pavesic, Max G. (1986). <u>Descriptive Archeology of Hells Canyon Creek Village</u>. Boise, ID: Boise State University, Archeological Reports, 14.

Pavesic's report covers five 1967 excavations at Hells Canyon Creek, just below the visitor parking lot north of the Hells Canyon Dam. The site was occupied by Native Americans for several hundred years. Pavesic was one of the pioneering archeologists in Hells Canyon.

Prose, Douglas (Producer). 1991. <u>Exotic Terrane</u>. Olney, PA: Bullfrog Films (Video—28 minutes).

This video tells the story of Hells Canyon and the Blue Mountains by using the Uniformitarianism Principle, which more or less states that "the present is the key to the past." The producer (and his associate producer, Nina Luttinger) shows that present day island arcs are good examples for the development of the Blue Mountains Island Arc. With outstanding graphics the video shows how plate tectonics processes work and how the Blue Mountains Island Arc was sutured or zippered to the more ancient North American Continent. I was the scientific advisor and worked closely with the producer. The video is an excellent companion to this book and can be purchased through Confluence Press and at any Forest Service visitor center in the Wallowa-Whitman National Forest.

Reid, Kenneth C. (Ed.). (1991). <u>An Overview of Cultural Resources in the Snake River Basin: Prehistory and Paleoenvironments</u>. Pullman, Washington State University, Center for Northwest Anthropology, 13.

Site-by-site descriptions of work in the Snake River Basin are given in this fine report. In particular, read the research history, geochronology, and cultural chronology sections (pages 22-55) and the section on Hells Canyon (pages 96-109). Dates and phases for the lower Snake region, beginning about 10,000 years ago, are given on page 36.

Reidel, S. P., Hooper, P. R., Webster, G. D., & Camp, V. E. (1992). <u>Geologic Map of Southeastern Asotin County, Washington</u>. Olympia, WA: Washington Division of Geology and Earth Resources, Map GM-40, Scale 1:48,000.

The authors compiled a geologic map of the area near the mouth of the Grande Ronde River. The map and explanation include the sequence of basalt formations (and flows) of the Columbia River Basalt Group, the pre-Tertiary rocks, and the Missoula Flood and Bonneville Flood gravel deposits.

Skow, John. (1967, July). Farewell to Hells Canyon. Saturday Evening Post, 76-83.
> This article is a requiem to Hells Canyon that was written in anticipation of the construction of High Mountain Sheep Dam. That dam would have flooded the canyon from a point near the confluence of the Salmon and Snake rivers nearly to the foot of Hells Canyon Dam.

Stratton, David H. (1984, March). Hells Canyon: the Missing Link in Pacific Northwest Regionalism. Idaho Yesterdays, 28, 3-9.
> This article is a succinct history of the anglos' exploration of Hells Canyon, beginning with the Lewis and Clark expedition in 1806.

The Wild and Scenic Snake River Boater's Guide Hells Canyon National Recreation Area. (1997). Washington, D. C.: U. S. Department of Agriculture, U. S. Forest Service.
> This book is a prerequisite for camping and boating in the canyon. It contains topographic maps of the canyon from Hells Canyon Dam to the Oregon-Washington border, in addition to campsites and regulations on boating and camping. Both geology and archeology are included where appropriate in order to enhance the visitors' experiences in the canyon. You can purchase the book at most U. S. Forest Service visitor centers in the region, particularly in Enterprise, Oregon, Clarkston, Washington, and Riggins, Idaho.

U. S. Army Corps of Engineers. (1988). Navigation Charts, Snake River, Oregon, Washington, and Idaho: Lewiston, Idaho to Johnson Bar. Walla Walla District, 14 sheets.
> This compilation contains some very fine navigation charts and is well worth the few dollars purchase price. Black and white photographs of the river corridor are included, along with river miles.

Vallier, Tracy L. (1967). The Geology of Part of the Snake River Canyon and Adjacent Areas in Northeastern Oregon and Western Idaho. Corvallis, OR: Oregon State University, Ph. D. Dissertation.
> This dissertation provides some of the first stratigraphic, petrographic, and structural information on the rocks between the Oxbow and Hells Canyon dams. Details on petrography, measured stratigraphic sections, and fossils are included.

Vallier, Tracy L. (1974). Preliminary Report on the Geology of Part of the Snake River Canyon. Portland, OR: Department of Geology and Mineral Industries, Map GMS-6, Scale 1:250,000.
> Map GMS-6 was the first attempt to combine the geology of Hells Canyon with some of the adjacent areas. You may want to compare GMS-6 map with those presented in this book to see the interpretive changes that I've made during the past two decades.

Vallier, Tracy L. (1977). The Permian and Triassic Seven Devils Group, Western Idaho and Northeastern Oregon. Washington, D. C.: U. S. Geological Survey Bulletin 1437, U. S. Government Printing Office.
> Bulletin 1437 describes the stratigraphy of the Seven Devils Group, which includes the Permian Windy Ridge and Hunsaker Creek formations and the Triassic Wild Sheep Creek and the Doyle Creek formations. This bulletin formalized those rock stratigraphic names with the Geologic Names Committee of the U. S. Geological Survey.

Vallier, Tracy L. (1994) Geologic Hazards in Hells Canyon, Eastern Oregon and Western Idaho. Denver, CO: U. S. Geological Survey Open-File Report 94-213.
> Geologic hazards in Hells Canyon include earthquakes, landslides, rock falls, floods, and debris flows, particularly those that cascade out of tributary canyons as blowouts. Man-made structures such as dams and mines also pose potential hazards. This report specifically discusses many of the hazards and points out some areas where they have occurred in the past and areas where they might be expected in the future.

Vallier, Tracy L. (1995). Petrology of Pre-Tertiary Igneous Rocks in the Blue Mountains Region of Oregon, Idaho, and Washington: Implications for the Geologic Evolution of a Complex Island Arc: In Geology of the Blue Mountains Region of Oregon, Idaho, and Washington: Petrology and Tectonic Evolution (pp. 125-209). Washington, D. C.: U. S. Geological Survey Professional Paper 1438, U. S. Government Printing Office.
> This article discusses the evolution of igneous rocks in three tectonostratigraphic terranes (Wallowa, Baker, and Olds Ferry) of eastern Oregon and western Idaho. The conclusions integrate the structure, stratigraphy, petrology, paleontology, radiometric dating, and paleomagnetism into an interpretation of the region's pre-Tertiary geologic evolution. Furthermore, I compared igneous rocks from the Blue Mountains Island Arc with rocks from the Aleutian and Tonga islands chains. The article forms the backbone for many of the interpretations presented in this book.

Vallier, T. L., & Brooks, H. C. (Eds.). (1986, 1987, 1993, and 1995). <u>Geology of the Blue Mountains Region of Oregon, Idaho, and Washington</u>. Washington, D. C.: U. S. Geological Survey Professional Papers 1435, 1436, 1438, and 1439, U. S. Government Printing Office.

These 4 professional papers were published in order to combine the many diverse research efforts on paleontology, petrology, stratigraphy, structure, and resources of the Blue Mountains region. Professional Paper 1436 (1987) contains chapters about the Idaho Batholith and the rocks around its borders. The separate chapters in the 4 books discuss the pre-Cenozoic rocks and were written for professional earth scientists. They will be, however, of interest to anyone who would like more specific geologic information about the region. Howard Brooks, my co-editor, has nearly 40 years' experience in the Blue Mountains, particularly in the Baker and Olds Ferry terranes.

Walker George W. (Ed.). (1990). <u>Geology of the Blue Mountains Region of Oregon, Idaho, and Washington: Cenozoic Geology of the Blue Mountains Region</u>: Washington, D. C.: U. S. Geological Survey Professional Paper 1437, U. S. Government Printing Office.

This U. S. Geological Survey professional paper complements the other 4 on the Blue Mountains that were edited by Vallier and Brooks. In 6 chapters, Cenozoic rocks of the Blue Mountains and adjacent regions are discussed. These chapters give the reader a better understanding of the Eocene, Oligocene, and Miocene strata and lava flows within the extensive Columbia Basin, mostly north and west of the Blue Mountains.

Wheeler, H. E., & Cook, E. F. (1954). Structural and Stratigraphic Significance of the Snake River Capture, Idaho-Oregon. <u>Journal of Geology, 62</u>, 525-536.

Wheeler and Cook discuss the origin of Hells Canyon, as it now appears, and suggest that it formed in large part during the draining of Lake Idaho. The authors speculate that about 2 Ma the Snake River was a tributary to the Salmon River.

White, D.L. (1972). <u>Geology of the Pittsburg Landing Area, Snake River Canyon, Oregon-Idaho</u>. Terre Haute, IN: Indiana State University, Master's Thesis.

White's thesis was the first detailed geologic study of the Pittsburg Landing area. His interpretations have been very important for later work in the region and are used extensively in this book.

White, David L., & Vallier, Tracy L. (1993) Geologic Evolution of the Pittsburg Landing area, Snake River Canyon, Oregon and Idaho. In <u>Geology of the Blue Mountains Region of Oregon, Idaho, and Washington: Stratigraphy and Resources</u> (pp. 55-73). Washington, D. C.: U. S. Geological Survey Professional Paper 1439, U. S. Government Printing Office.

This paper is a comprehensive summary of the geology of the Pittsburg Landing area including the stratigraphy, major fossils, and structure.

Glossary

Many of these definitions are directly quoted from, or slightly modified from, the fourth edition of *The Glossary of Geology*, published in 1997 by the American Geological Institute, 4220 King Street, Alexandria, Virginia 22302. Direct quotes from *The Glossary or Geology* are set in Roman type, whereas other definitions and explanations are given in *italics*. Where appropriate, I give examples of specific areas, rocks, and outcrops that help explain some of the terms.

A

accretionary prism. A generally wedge-shaped mass of tectonically deformed sediment at a subduction zone formed by the tectonic transfer of strata from the descending plate into the framework of the overlying plate. *Parts of the Baker Terrane were formed in an accretionary prism.*

alluvial fan. A low, relatively flat to gently sloping mass of loose rock material, shaped like an open fan or segment of a cone, deposited by a stream at the place where it issues from a narrow mountain valley upon a plain or broad valley. *Alluvial fans occur at the mouths of most tributary streams in Hells Canyon in the area below the Hells Canyon Dam.*

alum. A group of minerals containing hydrous aluminum sulfates. *The gossan near Willow Creek is referred to, by some, as an alum bed. Alum recovered from similar deposits elsewhere are reported to have medicinal values.*

ammonite. Any ammonoid belonging to the order Ammonitida, characterized by a thick, strongly ornamented shell with sutures having finely divided lobes and saddles. *Ammonites have been recovered from several formations in the Hells Canyon region. Ammonites are now extinct; a somewhat similar animal living today is the Pearly Nautilus.*

amphibole. A group of dark silicate minerals that includes hornblende. *Amphiboles are very common and are major rock-forming minerals in many igneous, metamorphic, and sedimentary rocks. In Hells Canyon, amphiboles are common in plutonic rocks like diorite and quartz diorite and in metamorphic rocks like amphibolite.*

amphibolite. A dark colored, mostly greenish black metamorphic rock that contains amphibole and plagioclase feldspar as major minerals. The minerals generally are aligned or lineated. *At least half the rocks in the Cougar Creek and Oxbow complexes are amphibolite. Most amphibolites in those rock bodies are deformed and metamorphosed basalt, diabase, and gabbro dikes.*

amphibolite facies. Metamorphic facies characterized by the minerals amphibole and andesine feldspar. *Most minerals in the amphibolite facies crystallize within a pressure range of 3 to 8 kbars and a temperature range of 450° to 750°C.*

andesite. A dark-colored and fine-grained flow rock containing andesine feldspar and one or more mafic minerals, generally either hornblende or a pyroxene (hypersthene or augite). This rock is very common in volcanic arcs and contains from 56 to 60 weight percent SiO_2. Extrusive equivalent of diorite.

angular unconformity. An unconformity that occurs between two groups of rocks whose bedding planes are not parallel or in which the older underlying rocks dip at a different angle (usually steeper) than the younger overlying strata. *In Hells Canyon, angular unconformities occur mainly between Triassic and Jurassic strata and between the pre-Tertiary rocks and the Miocene Columbia River Basalt Group.*

anticline. A fold with limbs that dip away from the core. The core contains older rocks than the limbs. The limbs are generally composed of bedded sedimentary rocks that dip away from an axis of the fold. *Several anticlines occur in the Martin Bridge Formation near Big Bar in Oregon and can be seen from the IPCO road. A nearly recumbent fold (horizontal axis) occurs in limestone of the Wild Sheep Creek Formation near the mouth of Cottonwood Creek.*

ash. Fine-grained pyroclastic material; also called volcanic ash. *Mazama Ash is the best example in Hells Canyon. Consolidated and lithified ash is referred to as tuff.*

B

basalt. A general term for dark-colored mafic igneous rocks composed chiefly of calcic plagioclase and clinopyroxene (augite). *Extrusive equivalent of gabbro. Basalt contains less than fifty-two weight percent SiO_2. Most basalt occurs in lava flows, dikes, and sills. Basalt and basaltic andesite are the most common flow rocks in the Triassic Wild Sheep Creek Formation in Hells Canyon. Andesite is minor. Basalt also comprises most of the Columbia River Basalt Group.*

basaltic andesite. A dark-colored extrusive rock with more silica than basalt, but with a similar mineralogy; it occupies the compositional niche (52-56 weight percent SiO_2) between basalt and andesite.

base level. The level of erosion below which a stream cannot erode its bed. *Each stream has a theoretical base level of erosion. Sea level is generally thought to be the ultimate base level. However, submarine currents can erode channels in even the deepest parts of the ocean.*

basement. The undifferentiated complex of rocks that underlies the rocks of interest in an area. *In Hells Canyon, the Cougar Creek Complex forms a basement to the stratified rocks. Beneath ocean floors, basalt lava flows generally form basement to the overlying sediments and sedimentary rocks. In places on continents, the oldest—generally Precambrian—rocks in an area are often referred to as the basement.*

batholith. A large, generally discordant plutonic body having an areal extent of forty square miles or more. *Batholiths are composed predominantly of medium- to coarse-grained igneous rocks (such as granite, granodiorite, and quartz diorite). The Idaho and Wallowa batholiths are examples; no batholith occurs in Hells Canyon.*

boudin. A tectonic lens that is elongate and shaped like a sausage. *Boudins are especially well exposed along the boundary of the plutonic body near Bills Creek and in dikes of the Trudy Mountain unit (informal name) in the Cougar Creek Complex.*

breccia. A coarse-grained clastic rock composed of large (greater than 2 mm in diameter), angular, and broken rock fragments that are cemented together in a finer grained matrix. Breccia is similar to conglomerate except that the fragments are angular rather than rounded; it may be sedimentary, pyroclastic, or tectonic (fault breccia). *Most sedimentary rocks in the Permian and Triassic Seven Devils Group are breccia and sandstone.*

C

caldera. A large, basin-shaped volcanic depression, more or less circular in form. *Calderas have large diameters, generally measured in miles. Best example in Oregon is the Crater Lake caldera that formed during the eruption of Mt. Mazama about 6,850 years ago. Calderas form by collapse of the volcano's top after magma is explosively erupted from the chamber beneath the volcano.*

cataclasite. A broken rock with fractured mineral grains; recrystallization is minor. *Upon recrystallization, however, a cataclasite may become mylonite. There are many cataclasites in the Oxbow Complex.*

celadonite. A soft, green or gray green, earthy, mineral of the mica group, generally occurring in cavities in basaltic rocks. A hydrous silicate of iron, magnesium, and potassium. *Bright green celadonite occurs in vesicles of lava flows within the Columbia River Basalt Group that crop out near the surge tanks of the Oxbow Dam.*

chlorite. A group of platy, green minerals that are widely distributed in metamorphic rocks; generally, they form as alteration products of ferromagnesian minerals (pyroxene, amphibole, and biotite) in igneous rocks. *Chlorite is one of the most common metamorphic minerals in rocks of the Permian and Triassic Seven Devils Group. Chlorite is common in rocks of the greenschist metamorphic facies.*

columnar joint. Structures that generally occur as parallel columns, either hexagonal or pentagonal in cross section, within both extrusive and intrusive rocks. The columns form by contraction during cooling. *Columnar joints are common in lava flows of the Columbia River Basalt Group. They also occur in Permian rhyolite dikes that crop out along the Snake River about one quarter of a mile south of its confluence with Sheep Creek. In places, like the cliffs near Asotin, Washington, the columns are curved and bent.*

conglomerate. A coarse-grained, clastic sedimentary rock composed of rounded to subangular fragments larger than two mm in diameter. *The larger fragments are set in a finer-grained matrix of sand, silt, and cementing materials. Clasts in conglomerates give geologists important information on source rocks and transportation mechanisms.*

crust (earth). The outermost layer or shell of the earth defined according to various criteria. *It refers to that part of the earth above the Mohorovicic discontinuity, which forms the acoustic—and also the compositional—boundary between the earth's crust and mantle.*

D

dacite. *A fine-grained extrusive igneous rock that occurs as lava flows, dikes, and sills. Composition lies between andesite and rhyolite. Extrusive equivalent of quartz diorite.*

debris flow. *A mass movement involving rapid flowage of debris of various kinds. Good examples are the high density mudflows generated by blowouts in tributary canyons after a heavy rain. Many of the volcanic breccias in the Wild Sheep Creek Formation were created by debris flows that cascaded off the slopes of underwater volcanoes.*

décollement. *Detachment structure of strata due to deformation. This term is generally associated with large-scale overthrusting. Décollements are common in subduction zones of island arcs where thick sediment layers on the downgoing oceanic crust slab are thrusted under overlying island arc or continental crust. A décollement exists under western Washington, western Oregon and northern California.*

detrital. *Formed from detritus, which is a collective term for loose rock and mineral material that is worn off or removed by mechanical means. The term detrital is often used to indicate a sediment source from outside a particular depositional basin.*

dextral. *Pertaining to something that grows or moves to the right. It can refer to coiling directions in ammonites, gastropods, and foraminifers; also refers to relative movement along strike-slip faults. The San Andreas fault in California has dextral (right-lateral) movement.*

diabase. *Dark-colored igneous rock, generally occurring in dikes and sills. It has the same composition as basalt, but is coarser grained and the same composition as gabbro, but is finer grained.*

dike. *A tabular igneous intrusion that cuts across the bedding or foliation of the country rock. The Cougar Creek and Oxbow complexes in Hells Canyon are composed mostly of dikes. Furthermore, dikes of the Columbia River Basalt Group are common along the slopes of Hells Canyon.*

diorite. *Intrusive equivalent of andesite with a coarse granitic texture. Common minerals are andesine feldspar and hornblende.*

discharge. *The rate of flow of surface water or ground water at a given moment, expressed as volume per unit of time. For example: cubic feet per second (cfs).*

E

epiclastic. *A term used to describe a sediment or sedimentary rock in which the fragments are derived by mechanical weathering and erosion; for example: conglomerate, sedimentary breccia, sandstone, and siltstone. Compare with pyroclastic.*

epidote. *A yellowish, pistachio-green silicate mineral with Ca, Fe, Al, and Si as major cations; commonly occurs in low-grade to medium-grade metamorphic rocks with albite and chlorite. A widespread mineral that results from the metamorphism of mafic rocks. Epidote is very common in rocks of the Seven Devils Group in Hells Canyon.*

exotic terrane. *A tectonostratigraphic terrane that has moved from its original position and is distinctly different from surrounding terranes. The Blue Mountains Island Arc and many other terranes in the Pacific Northwest of the United States, western Canada, and southeastern Alaska are exotic to the North American continent.*

F

facies. *An areally restricted part of a defined stratigraphic rock body, differing in lithologic character from other beds or strata deposited at the same time (for example, sedimentary facies). Also used to describe metamorphic mineral assemblages that crystallized within a fairly restricted temperature and pressure range (for example, greenschist and amphibolite facies).*

fault. *A fracture in rocks along which displacement occurred. Major fault types are normal, reverse, thrust, overthrust, strike-slip, and transform. Movement of faults commonly occurs during earthquakes.*

feldspar. Most common silicate mineral in the earth's crust; major cations are Ca, Na, K, Al, and Si; particularly widespread in igneous rocks. Plagioclase feldspars form a continuous series from the soda-rich variety (albite) to the calcium-rich variety (anorthite). Between the Na- and Ca-rich varieties are oligoclase, andesine, labradorite, and bytownite. Orthoclase is the most abundant potassium feldspar.

fissure eruption. An eruption that takes place from an elongate fissure rather than from a central vent. These are very common in basaltic eruptions that occur along oceanic spreading ridges like the Mid-Atlantic Ridge. Furthermore, most lava flows that compose the Columbia River Basalt Group were erupted from fissures.

flute. Shallow, oval-shaped hollows and grooves formed by a turbulent flow of water and commonly found in caves and along bedrock-lined banks of streams. A scalloped or rippled rock surface. The bedding planes of turbidite sandstones are often fluted.

fluvial. Of or pertaining to a river or rivers. Fluvial processes include the erosion, transportation, and deposition of materials by streams.

foliation. A general term for a planar arrangement of textural or structural features in any type of rock. Examples of mineral foliation are well displayed in gneiss, schist, mylonite, and gneissic mylonite. Many rocks in the Cougar Creek and Oxbow complexes are foliated.

footwall. The underlying side of a fault, orebody, or mine working. Compare with hanging wall.

formation. Basic rock-stratigraphic unit in a local or regional classification of sedimentary rocks; generally distinctive and mappable; important for interpreting the geologic history of a region. Two or more formations comprise a group. For example, the Windy Ridge, Hunsaker Creek, Wild Sheep Creek, and Doyle Creek formations compose the Seven Devils Group.

G

gabbro. Coarse-grained intrusive igneous rock with the same composition as basalt and diabase; can occur as dikes, sills, stocks, and batholiths. Essential minerals are labradorite feldspar and one or more pyroxenes, generally augite.

glomerol. Cluster of semi-equant phenocrysts in lava flows, sills, and dikes; texture is called glomeroporphyritic. Good examples occur in flows and sills of the Wild Sheep Creek Formation near Hells Canyon Dam.

gneiss. Foliated metamorphic rock, generally with alternating light- and dark-colored mineral bands.

gneissic mylonite. Metamorphic rock with gneissic foliation that formed locally along faults and other shear surfaces. Gneissic mylonites are common rocks in the Oxbow Complex.

goethite. A mineral that is generally yellow, but in places light rusty red; oxide with a chemical composition of $FeO(OH)$; common constituent of limonite; widespread in gossans of sulfide-bearing ore deposits. For example, abundant goethite occurs in the gossan near Willow Creek and is mainly responsible for the yellow, orange, and brown colors. Goethite is also common in the ore dump that occurs at the west portal (Imnaha River entrance) of the Mountain Chief Mine.

gossan. An iron-bearing weathered product overlying a sulfide deposit. It is formed by the oxidation of sulfides and the leaching-out of the sulfur and most metals, leaving hydrated iron oxides like goethite and rarely sulfates.

graded bed. A sedimentary bed, generally with coarser grained fragments at the bottom grading upward to finer grained fragments at the top. Graded beds are characteristic of turbidity current deposits. Graded beds are commonly referred to as turbidites.

greenschist facies. Low-grade metamorphic facies; the name is given to rocks containing an abundance of green minerals, generally epidote and chlorite. Pressures for the formation of greenschist-facies rocks are 3 to 6 kilobars and temperatures range from 300° to 500°C. Most rocks in the Seven Devils Group were metamorphosed to the greenschist facies.

H

hanging wall. The overlying side of an orebody, a fault, or mine working. Compare with footwall.

hematite. Brick red or black iron oxide mineral (Fe_2O_3); may be primary or secondary; principal ore of iron. Black specular hematite is common in the Mountain Chief Mine near the mouth of the Imnaha River.

hornblende. *Black or dark green silicate mineral; the most common mineral of the amphibole group; abundant in diorite, quartz diorite, and trondhjemite plutonic rocks and in the metamorphic rock, amphibolite.*

hot spot. A volcanic center, 100 to 200 km across and persistent for at least a few tens of millions of years, that is thought to be the surface expression of a persistent rising plume of hot mantle material. *The Hawaiian Island volcanoes form as the Pacific Plate moves over a hot spot. Volcanic rocks of the Snake River Plain probably formed over a hot spot that now lies under Yellowstone Park. A hot spot may have been responsible for the Triassic volcanic rocks in Wrangellia.*

I

intrusive. Pertaining to intrusion, both the process and the body so formed. *Rock bodies include dikes, sills, stocks, and batholiths along with a wide variety of other geometric bodies such as laccoliths and lopoliths; commonly referred to as a pluton.*

island arc. *A chain of islands rising from the sea floor, generally expressed as a curved line on the earth's surface. An oceanic volcanic arc. Examples are the Aleutian and Marianas island arcs. Not all island arcs are curved, but all rise above subduction zones in an oceanic setting and experience large earthquakes and abundant volcanic activity.*

K

K-Ar method. *Potassium-argon age method. Age determination of a rock or mineral in years, based on the known radioactive decay rate of potassium-40 to argon-40. Also called potassium-argon dating.*

keratophyre. *An andesite that has generally undergone metamorphism to the greenschist facies and is soda-rich. Albite is abundant. Chlorite and epidote may be common. See quartz keratophyre.*

L

landslide. A general term covering a wide variety of mass-movement landforms and processes involving the downslope transport, under gravitational influence, of soil and rock material en masse. *Many landslides have occurred in the past, and will occur again, in Hells Canyon and its tributary canyons.*

lithofacies. A lateral, mappable subdivision of a designated stratigraphic unit, distinguished from adjacent subdivisions on the basis of lithology. *For example, in places a rock body like limestone can be traced laterally into a sandstone body. These strata are the same age but formed in different environments of deposition. They are lithofacies.*

M

mafic. *Term used for dark-colored igneous rocks composed chiefly of ferromagnesian (Fe and Mg) minerals; term is also used for dark-colored minerals. Gabbro, diabase, and basalt are mafic igneous rocks, whereas olivine, pyroxene, and amphibole are mafic minerals.*

magma. Naturally occurring molten or partially molten rock material, generated within the earth and capable of intrusion and extrusion, from which igneous rocks are derived through solidification and related processes. *Magma forms in magma chambers and may be erupted onto the surface either as lava flows or pyroclastic material. The magma that remains under the surface will crystallize as a pluton (or intrusive).*

magnitude (earthquake). A measure of the strength of an earthquake, or the strain energy released by it, as determined by seismographic observations.

mantle (interior earth). The zone of the earth below the crust and above the core.

matrix. *Material filling the interstices between larger grains or particles in a sedimentary rock.*

metamorphism. The mineralogical, chemical, and structural adjustment of solid rock to new chemical and physical conditions which have been imposed at depth. *The heat for metamorphism may be generated near an igneous (generally intrusive) mass as it solidifies, during the movement of faults, and by the load of overlying rocks and thereby the earth's thermal gradient.*

mylonite. *A compact rock with a streaky or banded structure; generally forms by extreme deformation of pre-existing rocks, followed by recrystallization at depth. (See cataclasite and gneissic mylonite.) Mylonites are very abundant in the Oxbow and Cougar Creek complexes in Hells Canyon and formed during extreme deformation of pre-existing igneous rocks. Mylonite also is common along thrust faults, such as the one mapped south of Sheep Creek.*

N

norite. *Mafic intrusive rock similar to gabbro but containing some orthopyroxene (such as the mineral hypersthene) along with clinopyroxene (such as the mineral augite).*

P

paleomagnetic inclination. *Result, derived from paleomagnetic measurements, that shows the latitude of the rocks when they crystallized (igneous) or were deposited (sediments).*

paleomagnetism. *The study of natural remnant magnetism of earth materials in order to determine the intensity and direction of the earth's magnetic field in the geologic past. This technique was used to determine paleolatitudes of Permian and Triassic rocks in Hells Canyon and is in large part responsible for the interpretation that the rocks have moved a long distance since their formation.*

Pangea. *A hypothetical supercontinent that probably existed about 300 to 200 Ma. It supposedly combined all of the continental crust of the earth. The present continents were derived by fragmentation and movement away from each other. Pangea first fragmented into two large continents, Laurasia on the north and Gondwana on the south, which later broke up to form the continents we recognize today.*

pegmatite. *An exceptionally coarse-grained igneous rock, with interlocking crystals, usually found as irregular dikes, lenses, or veins. Quartz and feldspar are abundant, but many other minerals are unique to pegmatites. Pegmatite dikes cut Triassic gabbro and diorite between Wolf Creek and Dug Bar.*

phenocryst. *A relatively large crystal in a porphyritic igneous rock. Most lava flows in the Seven Devils Group are porphyritic.*

pillow lava. *A general term for those lavas displaying pillow structure and considered to have formed in a subaqueous environment. Pillow lavas and pillow breccias are very common in outcrops north of Pittsburg Landing in the Big Canyon unit of the Wild Sheep Creek Formation.*

pluton. *An igneous intrusion, generally referring to rock bodies larger than dikes and sills.*

plutonic. *Pertaining to igneous rocks formed at great depth, most commonly stocks and batholiths (see intrusive).*

porphyritic. *Texture of igneous rocks; large crystals (see phenocryst) are set in a finer groundmass which may be glassy or crystalline. This texture implies an interrupted crystallization history.*

pyroclastic. *Pertaining to clastic rock material formed by volcanic explosion or aerial expulsion from a volcanic vent; also pertains to rock texture of explosive origin. Compare with epiclastic. Common adjective is tuffaceous.*

pyroclastic breccia. *Refers to pyroclastic fragmental debris with an average grain size greater than about 4 mm. Finer-grained pyroclastic material is called ash if unconsolidated and tuff if lithified.*

pyroxene. *Dark, rock-forming silicate mineral that is widespread in mafic igneous rocks; particularly common in basalt, diabase, and gabbro. Prevalent pyroxenes are augite and hypersthene.*

Q

quartz. *Crystalline silica (SiO_2); next to feldspar, quartz is the most common mineral in the earth's continental crust.*

quartz diorite. *A coarse-grained plutonic rock more silicic than diorite; free quartz is common.*

quartz keratophyre. *Sodium-rich rhyolite common in older strata of some island arcs. Large bipyramidal quartz phenocrysts may be common. The Hunsaker Creek Formation contains quartz keratophyre dikes, sills, tuffs, and rare lava flows. Generally, but not necessarily, quartz keratophyre is the result of low-grade metamorphism.*

quartzofeldspathic. *Said of rocks that are rich in quartz and feldspar.*

R

rhyolite. *Extrusive, generally porphyritic, igneous rock that may exhibit flow texture. The rock often contains quartz and feldspar phenocrysts. The rocks contain high amounts of SiO_2 (greater than 70 percent). In Hells Canyon rhyolites in the Hunsaker Creek Formation are sodium-rich and potassium-poor (see quartz keratophyre).*

S

schist. *A strongly foliated crystalline rock, formed by dynamic metamorphism, that can be readily split into thin flakes or slabs due to the well developed parallelism of the minerals present. Protolith may be either sedimentary or igneous. Schists are uncommon in the Hells Canyon region, but are very abundant in rocks of the Riggins Group that crop out along the Salmon River near Riggins, Idaho.*

silicic. *Term often applied to igneous rocks and magmas that are silica-rich (greater than about 65 percent).*

sill. *A tabular igneous intrusion that parallels the bedding or foliation of the sedimentary and metamorphic country rock, respectively. Compare with dike.*

sinistral. *To the left. Applies in this book to left-lateral movement along strike-slip faults. Compare with dextral.*

sinkhole. *Closed depression in an area of karst topography, generally formed by collapse of an underlying cave.*

slump. *A type of mass movement that may lead to landsliding. Generally characterized by a curved headwall that is pulled away from coherent rocks or sediments. The best example of a slump in Hells Canyon occurs between Marks and Waterspout creeks.*

stock. *An igneous intrusion (pluton) that has less than 40 square miles of surface exposure; resembles a batholith except in size. Most of the larger plutonic bodies in Hells Canyon are stocks.*

strike-slip fault. *A fault that has movement parallel to the strike (trend) of the fault.*

subaerial. *Said of conditions and processes, such as erosion, that exist or operate in the open air on or immediately adjacent to the land surface, or of features and materials that are formed or situated on the land surface.*

subduction zone. *An elongate region along which a crustal block descends beneath another crustal block. The Baker terrane formed in part along a subduction zone. A subduction zone may include an accretionary prism. Characteristic of the fore-arc region of island arcs. Subduction zones occur along the west coasts of South and Central America, Mexico, the Pacific Northwest, and southern Alaska.*

T

tailings. *Rocks dumped at the portals of mines and prospects. Tailings are composed of rocks and minerals that are not worth milling and smelting.*

terrace. *Any long, narrow, relatively level or gently inclined surface, generally less broad than a plain, bounded along one edge by a steeper descending slope and along the other by a steeper ascending slope. Terraces generally occur above the level of a body of water and along its margin. Best examples in Hells Canyon occur upstream from Temperance Creek and in the Johnson Bar area.*

tectonostratigraphic terrane. *A fault-bounded rock body or a mixture of rock bodies of regional extent characterized by a geologic history distinct from that of neighboring terranes. See exotic terrane.*

thrust fault. *A fault with a dip of forty five degrees or less over much of its extent, on which the hanging wall appears to have moved upward relative to the footwall.*

transform fault. *A strike-slip fault characteristic of mid-ocean ridges and along which the ridges are offset. Analysis of transform faults is based on sea-floor spreading processes.*

transgression. *The spread or extension of the sea over land areas. This generally occurs during a world-wide rise in sea level or subsidence of a landmass. A transgression occurred while the Coon Hollow Formation was being deposited.*

trondhjemite. *Quartz-rich quartz diorite; contains more than 30 percent quartz. The plutonic rock that crops out along the west side of the Oxbow Dam is trondhjemite.*

tuff. *A compacted pyroclastic deposit of volcanic clasts and volcanic ash. Many rocks in the Hunsaker Creek Formation are submarine tuffs, meaning that they were deposited under water. Some of the pyroclastic eruptions may have actually occurred under water.*

turbidity current. *A bottom-flowing current laden with suspended sediment that moves rapidly down a slope and spreads horizontally across the floor of an ocean or lake.*

turbidite. *Sediment deposited from a turbidity current; it generally consists of sand and silt and is characterized by graded bedding.*

U

unconformity. *A substantial break or gap in the geologic record where a rock unit is overlain by another that is not next in stratigraphic succession. It generally represents a period of erosion. See angular unconformity.*

V

volcaniclastic. *Rock containing fragments of volcanic rocks and minerals; may be pyroclastic or epiclastic.*

W

waterspout. *Throughout the American West a waterspout describes a flash flood that occurs in a narrow canyon; also referred to as a stream blowout. Results are debris flows that form alluvial fans and temporary dams.*

wind gap. *A former stream channel, now abandoned by the stream that formed it. Wind now whistles through the gap where water once flowed. There are Bonneville Flood water gaps at the Oxbow and near Upper Pittsburg Landing.*

X

xenolith. *An inclusion in an igneous rock to which it is not genetically related.*

Care of the Canyon and Hiking Safety

The first and foremost advice for taking care of the canyon and for hiking safety is to use common sense and to plan for the unexpected.

The rock art (petroglyphs and pictographs) left by Native Americans cannot be replaced and is an important part of the region's history. Take pictures, but don't touch.

Leave arrowheads and other artifacts undisturbed. You may make a very important scientific discovery. If you discover an artifact, mark the site with a rock cairn. Make a map with directions and approximate distances and turn the information over to Forest Service personnel. Treat fossils with the same respect. There are only a few fossil sites in Hells Canyon. Each is precious.

Hiking and camping in the Hells Canyon and Seven Devils Mountains regions can be very rewarding. The trails that parallel the river are relatively easy to hike, and they are well maintained. However, if you decide to hike other trails or to walk cross country, plan and prepare your trek carefully. Based on my experiences in the canyon, I offer the following advice. Keep in mind that this list is not all-inclusive.

- Wear good hiking boots. I prefer leather boots, at least eight inches high and with thick soles, for protection from snake bites, cactus, rock scrapes, and twisted ankles. Running shoes and ankle-high boots of man-made materials and soft leather may be adequate for short hikes along canyon-edge trails, but definitely not for cross-country hikes.

- Wear long pants and take along a shirt with long sleeves. Poison ivy (also called poison oak) abounds along the trails as does Prickly Pear cactus at lower elevations. Evenings can be chilly at upper elevations, even in the middle of summer, and sudden rains can lower the temperature many degrees in a short time interval. Carry a jacket or coat.

- To avoid blisters, consider wearing two pair of socks. Thin rayon or nylon socks can be worn under heavy wool or cotton socks. The thin socks will more or less become molded to your feet. The other pair will slide back and forth over the thin socks rather than over the skin of your feet. Take off the boots and socks frequently to check your feet, particularly if they were not toughened from previous hikes. Change socks often if they are wet.

- Be alert for rattlesnakes, particularly in the morning and throughout the day in shaded areas. Learn about treating snake bites before you enter the canyon. If bitten, don't panic. Get to a doctor as soon as possible. Most people survive a rattlesnake bite, but it isn't a pleasant experience. Children are in greater danger from a bite than adults.

- Always take along plenty of water and a hat. Sunstroke is a hazard during the summer months. It is best to carry your own drinking water that was taken from an approved source rather than to rely on water from creeks. If you use creek or river water, either filter, treat with chemicals, or boil before drinking. I carry a small bottle of chlorox (three to four drops per quart) or iodine pills to treat water in an emergency.

- Avoid hiking alone. Be sure a responsible party knows where you will be at certain times. If you are taking a long hike, think about what you will do if forced to remain away from camp for one or more nights. Throw a jacket into your pack. Take along trail mix or other light-weight food that can last you for at least 48 hours. Carry a whistle or other signaling device, and agree upon a distress signal (such as three shrill whistles if you need help). A mirror can be used to signal boats, airplanes, and helicopters. Carry a flashlight. Don't take short cuts. Carry a USGS topographic map of the area and a compass. An altimeter is very useful if you have a topographic map. Global Positioning System (GPS) equipment is relatively inexpensive. Many hikers now carry GPS equipment routinely. If you are lost, it is best to remain on the trail and follow it toward the river. Wait near the river for a passing boat.

•Carry a first-aid kit and know how to use it. A small wilderness first-aid book is very handy. Always keep some sun block lotion with you.

•If you go cross country, allow at least twice the time that a hike would take on relatively easy trails like those along the river. The seemingly short distances between points on a map can easily fool an inexperienced hiker in the rugged terrain of the canyon.

•Assume that all animals are dangerous. Wild predators are becoming more common in the canyon and some are very territorial. In particular, don't get between a mother bear and her cubs. I have seen cougars, bears, lynxes, bobcats, and coyotes in the canyon. Clean up the campground each night, and safely stow away all food before retiring. Otherwise, you may have an unwelcome visitor.

•Expect obstacles along and on the trail. You may be scrambling over fallen trees or skirting a landslide.

•Almost all creeks have the potential to experience a blowout. When it is raining, and early in the summer or late spring season when snow is melting, be extremely careful where you camp. Camp at a distance from, and above, the stream channel and on higher levels of alluvial fans. Furthermore, don't camp below steep cliffs where rockfalls may occur.

•Be cautious with fire. The canyon is closed to open fires throughout most of the summer. Check with Forest Service personnel about rules for campfires. A backpack stove that uses a small propane tank is an excellent substitute. Better yet, take along food that doesn't require a stove or fire for its preparation.

•Human waste is a problem in the canyon, particularly along rivers and streams. Talk to a U. S. Forest Service ranger about depth of waste burial, disposal of toilet paper and sanitary napkins, and distances away from streams and trails that are considered appropriate.

•The canyon has a delicate ecosystem. Humans can easily upset parts of the ecosystem just by their presence. The natural balance can be upset by killing one rattlesnake or even one spider. Barrel cacti are extremely rare and should not be damaged; I've only seen a few during all my years in the canyon. Respect the animal and plant dwellers in the canyon. They are residents. We are visitors.

Acknowledgments

I have had the opportunity to work with many exceptional people during the more than three and a half decades that I've been involved with studies in the Hells Canyon region. The following acknowledgments include many of them. I also list the institutions and agencies that supported, either directly or indirectly, my studies in Hells Canyon.

First and foremost I gratefully acknowledge the patience and assistance of family members. Trudy, my wife for more than 32 years and now deceased, sacrificed to ensure that I had the funds and time to follow my research interests, encouraged me at every step, and spent several field seasons in the Hells Canyon region, providing logistical support and other assistance. My present wife, Sheila, understands my scientific and personal pursuits and encourages their successful completion. Sons Garry, Lane, and Monte and daughter, Susan, spent several long summers in the Hells Canyon region and assisted in many ways. They missed baseball and soccer games and time with their friends. My brother Kent was a capable assistant during two field seasons in the early 1990s. His cooking skills, sense of humor, and dedication to the mapping project are greatly appreciated. He also commented on an early draft of this book. My sister, Shirley Remmenga, and her husband, Elmer, worked with me in several rugged parts of Hells Canyon. My niece, Linda Fisher, and nephew, Donald Angeroth, provided several weeks of field assistance.

Bill Taubeneck, major professor and mentor, introduced me to the geology of eastern Oregon and directed my graduate studies at Oregon State University. He is an expert on granitic rocks in eastern Oregon and an outstanding field geologist. Without his early advice, and later on his encouragement, I would not have worked in Hells Canyon.

Howard Brooks taught me the geology of the Baker and Olds Ferry terranes and is a valued scientific colleague and friend. His encouragement strengthened my resolve to continue studies in the Blue Mountains at times when I was broke and discouraged. He and I co-authored several articles and, together, we compiled and edited 4 books on the geology of the Blue Mountains region and the Idaho Batholith (U.S. Geological Survey Professional Papers 1435, 1436, 1438, and 1439).

David White was my field assistant in 1968 and my first geology graduate student (M.S. in Geology, Indiana State University). He is a respected scientific colleague and friend. Dave continually challenges my interpretations and has made several significant contributions to our understanding of Hells Canyon geology. He reviewed 2 early drafts of this manuscript and provided many helpful comments and suggestions.

I am particularly grateful to David Fredley, my field assistant in 1970, and a valued colleague for more than 25 years. Dave is an ardent supporter of my work and provided monetary assistance and other support through the U. S. Forest Service. Without his help, this book would not be possible.

Ellen Bishop shared her knowledge and enthusiasm for the Blue Mountains region with me, beginning in the late 1970s. She put data and fresh interpretations into some poorly supported speculations that had been made about the region. Ellen reviewed early drafts of this manuscript and provided helpful comments and suggestions.

David Scholl encouraged my research efforts in island arcs and accompanied me on several research cruises to the Aleutian and Tonga island chains. He has been a valued colleague, a challenging mentor, and a good friend for 25 years. Dave, as much as anyone, believes that studies of sea floor rocks and sediments are important for understanding rocks on land.

I greatly appreciate the help provided by Nicholas Walker who determined important radiometric ages for plutonic rocks in the Blue Mountain region using uranium-lead methods. We worked together during the 1979 field season. His data were crucial for understanding the diverse assortment of plutonic rocks in Hells Canyon.

George Stanley, Sidney Ash, Norm Silberling, Cathy Newton, Ralph Imlay (deceased), and Frank Stehli provided significant paleontological results. Without their help there would be no stratigraphic column for the rocks.

Doug Prose and Nina Luttinger produced the video, *Exotic Terrane,* that tells a geologic story of the Blue Mountains Island Arc and its collision with ancient North America. I greatly appreciate their friendship and skills.

John Byrne taught me about the ocean floor and changed my professional life by recommending that I go to sea as a scientist on the Deep Sea Drilling vessel, GLOMAR CHALLENGER, in 1969 (Leg 5). The Deep Sea Drilling Project (now called the Ocean Drilling Program) has been the most successful earth science program in human history. It provided the data that led to the confirmation of the plate tectonic hypothesis.

James Hepworth edited and published this book through Confluence Press. I am grateful for his editing skills, patience, and encouragement.

Susan Schroeder (deceased) introduced me to James Hepworth at Confluence Press and reviewed the first draft of this book. Her enthusiasm and encouragement for this book are gratefully acknowledged. She and her husband, Ned, taught me how to raft through the rapids in Hells Canyon.

I appreciate the comments and suggestions that were made by Sheryl Grant on parts of the manuscript.

Field assistants who provided help for 2 or more weeks, besides those mentioned above, include Jim Waldrip, Ron Ozier, Julie Swank, and Chris Gibson. Rob O'Connor accompanied me in the Aleutian Islands and in Hells Canyon. Many others accompanied me in the canyon for shorter periods of time and I thank them for their assistance. Gary Mann worked diligently with me in the Cuddy Mountains region of western Idaho. His insights greatly helped my interpretations of Hells Canyon geology. I thank Bill Harbert, Jack Hillhouse, and Sherm Grommé for their studies of paleomagnetism.

In addition to David Fredley (now retired), I thank other employees of the U. S. Forest Service for support and assistance, including Art Seamans (retired), Lynn Sprague, Woody Fine, Roy Lombardo, Bruce Womack, Ed Cole, Jane Rohling, Tom King, Mike Cole, Earl Baumgartner, and Pat Worle.

I thank all of my former supervisors at the U.S. Geological Survey who supported my continued work in Hells Canyon, often when it was outside my assigned duties. Hollis Dole (deceased) and Roland Reid supplied monetary assistance from the states of Oregon (1965) and Idaho (1968) at critical times during my studies. Mark Ferns freely shared with me his knowledge of the Baker terrane and of regional mineralization and has encouraged my continued involvement in geologic studies of Oregon.

Several universities, scientific societies, and state and federal agencies supported my studies with monetary and other assistance. These include Oregon State University, Indiana State University, University of California at San Diego, Lewis-Clark State College, Geological Society of America, Society of Sigma Xi, Oregon Department of Geology and Mineral Industries, and the Idaho Bureau of Mines and Geology (Idaho Geological Survey). The U.S. Geological Survey and U.S. Forest Service provided some funding for this work.

I appreciate the encouragement and assistance of my colleagues at Lewis-Clark State College, particularly Scott Linneman, Mike Vernon, Tom Urquhart, Bill Laval (retired), Bill Heins, and Jeff Matthews. Scott reviewed an early draft of this book.

And last, but not least, I thank my many friends and acquaintances in Oxbow, Halfway, Baker City, and Enterprise, Oregon, plus those in Lewiston and Riggins, Idaho, and in Clarkston, Washington. I gratefully acknowledge, in particular, the help and interest of Floyd Harvey, George and Lynette Hauptman, Gary Armacost (deceased), Brett and Doris Armacost, Dixie Taylor, Mike and Jodee Luther, and Wally and Myrna Beamer.